Reena Chopra

# Wild Mammals

# Wild Mammals

## of Northwest America

Arthur and
Candace Savage

The Johns Hopkins University Press, Baltimore and London

Printed in Canada by Modern Press
Saskatoon, Canada.

This edition authorized by Western Producer Prairie Books, Saskatoon, Canada. Published simultaneously by The Johns Hopkins University Press, Baltimore, Maryland 21218, and the Johns Hopkins University Press Ltd., London.

Book and cover design by John Luckhurst
Warren Clark Graphics Ltd.

Library of Congress Cataloging in Publication Data

Savage, Arthur.
Wild mammals of northwest America.

Bibliography: p. 177
Includes index.
1. Mammals - Pacific, Northwest. I. Savage, Candace Sherk, 1949- . II. Title.
QL719.N93S28 599.09795 81-47589
ISBN 0-8018-2627-6 AACR2

Library of Congress Cataloging in Publication Data

Savage, Arthur.
Wild mammals of northwest America.

Bibliography: p. 177
Includes index.
1. Mammals - Pacific, Northwest. I. Savage, Candace Sherk, 1949- . II. Title.
QL719.N93S28 599.09795 81-47589
ISBN 0-8018-2627-6 AACR2

For Diana

# Contents

Scientific Advisors ix

Introduction 1

Order Insectivora
Shrews and Moles 5

Order Chiroptera
Bats 8

Order Lagomorpha
Pikas, Rabbits, and Hares 13

Order Rodentia
Rodents 22

Orders Odontoceti and Mysticeti
Whales, Dolphins, and Porpoises 59

Order Carnivora
Carnivores 72

Order Pinnipedia
Sea Lions and Seals 131

Order Artiodactyla
Cloven-hoofed Mammals 139

Checklist 175

References 177

Picture Credits 201

Index 203

# Scientific Advisors

The authors gratefully acknowledge the assistance of the following experts, each of whom read and commented on one or more sections of the manuscript, as indicated below:

Ian McTaggart Cowan, PhD, Honorary Professor, Department of Zoology, University of British Columbia — carnivores

David R. Gray, PhD, Assistant Curator, Vertebrate Ethology Section, National Museum of Natural Sciences — muskox

Valerius Geist, PhD, Professor and Associate Dean, Faculty of Environmental Design, University of Calgary — cloven-hoofed mammals

J. David Henry, PhD, field biologist and author — canids, black bear, and grizzly bear

Stefani Hewlett, BA, Staff Biologist, Vancouver Public Aquarium — sea otter; sea lions and seals; whales, dolphins, and porpoises

Karl W. Kenyon, MSc, Wildlife Biologist, Research, United States Fish and Wildlife Service (retired) — sea otter; sea lions and seals; whales, dolphins, and porpoises

Robert E. Wrigley, PhD, Museum Director and Curator of Mammals and Birds, Manitoba Museum of Man and Nature — introduction; moles and shrews; pikas, rabbits, and hares; rodents

Ultimate responsibility for the accuracy of the text lies with the authors.

# Introduction

Northwest America is home to an astonishing variety of wild mammals. This book deals with more than seventy representative species, and every one of them is interesting. Some must surely number amongst the oddest creatures on earth — the uncuddly porcupine with its armor of quills, the boldly striped skunk with its reeking spray, the gangly moose with its mournful-looking face. Others are awesome, both in reputation and in fact, such as the cougar and the grizzly bear. Many, like the squirrels, foxes, and whales, hold a special appeal for humankind. Even the less "lovable" species, such as the humble meadow vole, gain in stature as one begins to appreciate the sophisticated mechanisms and behaviors that regulate their lives.

One of the most impressive characteristics of these animals, taken as a group, is their diversity. In size, they range from the lightest of the shrews, at three grams or less, to certain whales, at a hundred tonnes and more. For locomotion, they may have paws or hooves, flippers or flukes; some of them even have wings. They are active by day or night or at twilight, in the air, on land, in the water, in burrows underground. Some, like the pronghorn, are fleet and graceful; others, like the star-nosed mole, seem bumbling and grotesque. Yet each is admirable in the intricacy with which it is adapted to its specific habitat.

The variety exhibited by these mammals is a result, in part, of the variety of life zones to be found in northwest America. For our purposes, northwest America has been defined as western Canada, together with the neighboring province and states. Since wild animals are not much impressed by political boundaries, most of what is reported here applies throughout this area. But the timing of certain natural events, such as mating and hibernation, may vary somewhat with latitude. In these cases (and wherever else it was felt that the text could benefit from a more narrowly specified geographic range) western Canada was chosen as the area of primary focus.

To describe or even to enumerate the major types of habitat that occur in the area with which we are concerned is no easy task. In the far North lies an inhospitable, stony desert that grades into heath and then into scrublands as one proceeds south; together these form the tundra. Then comes the northern forest, with stands of lichen-encrusted spruce, broken occasionally by pine, fir, or tamarack, aspen or birch. Further south is the parkland transition, where deciduous forest opens to patches of meadow and then cedes to true grasslands. Far to the west, in the coastal "wet belt," one encounters the almost tropical lushness of the fir-and-cedar forests. And finally, proceeding up the slopes of the mountains, there is a gradual return to the tundra and the barrens of rock and ice. Within each of these principal zones are countless subzones, or microhabitats, each with its characteristic forms of life, including mammals. This should be borne in mind when using the distribution maps that are found throughout the book: within the ranges indicated, each species can be found only where its particular microhabitat occurs.

For all their variability in appearance, range, and life history, the animals treated in this book have much in common. In the first place, as mammals, all of them are warm-blooded. That is to say, like birds, they are capable of maintaining a relatively high and constant body temperature. This is in contrast to cold-blooded animals, such as amphibians and reptiles, in which body temperature — and hence the speed of metabolic reactions — depend on the temperature of the environment. It is because mammals are warm-blooded that they are able to maintain high levels of physiological and physical activity, a characteristic that does much to account for their success as a class.

The animals in this book also share two

**Of the many dozen species of wild mammals in our region, one of the most impressive is the mountain lion. Yet even the "humbler" species acquire a certain magnificence as one comes to know them well.**

uniquely mammalian characteristics: hair and mammary glands. Any species that possess hair, however sparse, and mammary glands, however primitive, can be unequivocally classified as mammals. (Hence, people clearly number amongst those that qualify.) Interestingly, both these adaptations are related to warm-bloodedness. In the case of hair, the link is obvious: it acts as insulation to reduce the loss of body warmth to the environment. But mammary glands serve to produce milk for the nourishment of young. What do they have to do with thermoregulation? The answer lies in their history, since they are thought to have evolved out of sweat glands, which function as cooling organs. Thus, had it not been for the development of the means to regulate body temperature, mammals would probably not have acquired the capacity to suckle their offspring.

Various other mammalian characteristics can also be traced back to their warm-bloodedness. For example, many (though not all) mammals have four kinds of specialized teeth. At the front are incisors for nipping; flanking them on both sides are canines for piercing; and farther back are premolars and molars for crushing and grinding. Together these increase the animals' ability to obtain and utilize food, thereby helping them to meet the high energy requirements that go along with being warm-blooded. They also have four-chambered hearts, a feature that is found in birds as well. By preventing "used" venous blood from mixing with the freshly oxygenated arterial supply, this adaptation increases the amount of oxygen that is available in the cells to fuel metabolism. In addition, mammals, and only mammals, are equipped with muscular diaphragms, which increase the air-holding capacity of the chest cavity.

This improved oxygen supply, which nourishes the brain as well as other tissues, is partly responsible for the superior intelligence of mammals. Generally speaking, mammals have more brain per gram of body weight than do other animals, an increase that is principally due to the addition of "gray matter" — the tissue that is associated with many aspects of memory and learning. These mental powers permit mammals to profit from experience to a unique degree. Because of the period of training that often accompanies nursing, the young may even be able to benefit from the discoveries of past generations. Thus, the behavior of mammals is more variable than that of any other class of animals, because it is more often guided by intelligence.

In these and many other ways, mammals have far outstripped their ancestors. The exact evolutionary origins of mammalian characteristics will remain obscure forever, because details of the soft anatomy are not often preserved in fossils. But through studies of fossil bones and teeth, many facts about mammalian ancestors have literally been unearthed. Mammals all have their origins in the therapsids, primitive, mammal-like reptiles that arose not near the end of the Age of Reptiles, as one might suppose, but very early in reptilian evolution, about 250,000,000 years ago. The first true mammals were small, mouse- or rat-sized creatures, generally shy and nocturnal, that lived mainly on a diet of insects. During the long ages when dinosaurs were the masters of the Earth, these primitive mammals played an inconspicuous role, pursuing their own course of development. When the Age of Reptiles came abruptly and mysteriously to an end, about 70,000,000 years ago, the mammals emerged as the dominant terrestrial animals. In the ensuing Age of Mammals, they flourished and diversified, filling thousands of evolutionary niches with their multitudinous forms.

Most of the mammals that occur today in North America immigrated relatively recently from Asia, via an intermittent land bridge in what is now the Bering Strait. Many of the newcomers arrived within the last 3,000,000 years, during a period when a series of vast glaciations, or Ice Ages, caused ocean levels to fluctuate, alternately exposing and inundating the bridge. Among the most recent immigrants were Stone Age people, who first arrived perhaps around 100,000 years ago. Masterful hunters, equipped with sophisticated stone-tipped weapons and skilled in the use of fire to drive game, the Paleo-Indians may have had an enormous impact on the mammals of the New World. In a few thousand years, one brief moment of geological time, a fantastic variety of animals vanished. Mastodons, mammoths, giant beavers, ground sloths, giant armadilloes, horses, tapirs, camels, and saber-toothed cats were some of the casualties that may have accompanied the establishment of primitive people in North America.

The study of extinct mammals, especially the more ancient ones, is a gold mine of information for the taxonomist. By combining a knowledge of the fossil record with comparative studies of living types, one can classify mammals not only to provide unambiguous scientific names but also to show evolutionary relationships. This is accomplished by grouping the mammals into a series of ranks, or taxons, each of which includes all those types that are thought to have a common ancestry. For example, the class — in this case Mammalia — includes all mammals: they are united by their common ancestry amongst the therapsid reptiles. Progessively finer subdivisions in the taxonomic hierarchy — order, family, genus, and species — include mammals that, in general, have more recent common ancestors. The basic unit of classification is the species, which is loosely defined as any group of animals capable of interbreeding freely and producing fertile offspring. Black bears, for example, form a species: they interbreed amongst themselves but normally do not mate with any other animals, not even their close relatives, the grizzlies. Both the distinctness and similarity of these two species are reflected in their scientific names: *Ursus americanus* for the black bear and *Ursus arctos* for the grizzly. The species names, *americanus* and *arctos,*

**Anatomy is not all that is affected by natural selection; behaviors also evolve. This arctic ground squirrel, for example, has the innate ability to warn its neighbors about the approach of particular kinds of predators, by uttering the appropriate call.**

differ, but the genus name, *Ursus,* is the same. Sometimes a third word is added to the scientific name to denote a particular subspecies, or geographic variant.

The anatomical differences that have evolved amongst mammals are the basis for their classification, but many behavior patterns, too, are naturally selected characteristics. While mammals can rapidly learn to modify their actions through experience, this ability is superimposed on a variety of innate, or unlearned, behaviors, which vary from species to species. Particularly intriguing is social behavior, the way animals relate to others of their own kind. Most mammals are not highly sociable or gregarious and, except for mating and rearing young, lead mainly solitary lives. Interactions within such a species are apt to be hostile because there is competition for food, shelter, and other necessities that are available in limited quantities for any animal filling a particular ecological role, or "niche," in the community. This commonly leads each individual to establish a territory — a portion of its home range (or usual sphere of activity) that is staunchly defended, by scent-marking, threats, or overt violence, from other members of the same species.

The available resources may also be partitioned by dominance hierarchies (sometimes known as peck orders, since they were first described in birds). These social systems are most obvious in the more gregarious species such as wolves and mountain sheep. Each member of such a group knows which of its companions it can dominate and to which it must submit. Social position is all important, for it often regulates access to food and to bedding or denning sites and frequently determines which individuals will have the chance to mate and reproduce. Rank is generally determined by fighting, but savage, fatal encounters are uncommon; instead, ritualized contests, in which neither animal runs much risk of serious injury, commonly determine which of two adversaries will be dominant.

Such behaviors as the defense of territories and the establishment of hierarchies have evolved because, in one way or another, they increase the species' capacity to survive. The process of natural selection, through which these adaptations have arisen, is both subtle and complex. Yet for many species of mammals — including some of those discussed in this book — natural mechanisms will not be sufficient to ensure their survival. It is difficult to think of a single native species that has not been affected somehow, for good or ill, by the human onslaught of the last century. Habitat has been altered on a massive scale, contributing to the virtual extinction of some species and to the proliferation of others. Unrestrained hunting, trapping, and whaling have periodically brought about severe population declines among victimized mammals. A number of species, such as the Vancouver Island marmot, the sea otter, and the humpback whale, are currently in the danger zone of low or declining populations. If they are to endure, these and other species will require the well-informed cooperation of mankind.

This book aims, first, to inform its readers, by providing an accurate, nontechnical survey of the most authoritative popular and scientific literature. Beyond that, the text is designed to be interesting and full of surprises, with reports on many intriguing aspects of behavior and ecology. And ultimately, the book is intended to foster a widespread and informed interest in the survival of wild mammals. In many ways, their well-being is our own.

## Order Insectivora

# Shrews and Moles

The order Insectivora is a grab bag of small placental mammals, including hedgehogs, shrews, moles, and several lesser-known groups. Comparatively speaking, they are unspecialized and rather primitive animals, not very different in many ways from mammals that lived 100,000,000 years ago. In western Canada, this ancient order is represented today by thirteen species in all. One of them is the star-nosed mole, *Condylura cristata,* which occurs in southern Manitoba and regions to the east and south. It is a weird-looking creature that lives largely underground, feeling its way with a cluster of sensitive tentacles that festoon the end of its snout. There are three other species of moles in western Canada, all of which are confined to southwestern British Columbia and the coastal states to the south; the other nine species of insectivores in our area are all shrews.

The masked shrew, *Sorex cinereus,* is one of the most common and widely distributed of our mammals. Yet chances are good that you have never seen one. There are several reasons for this, the most obvious being that shrews are very small. The short-tailed shrew, *Blarina brevicauda,* which is our largest species, usually weighs just over twenty grams, or about two-thirds as much as an average meadow vole. A good-sized masked shrew tops the scale at around four grams, somewhat less than a twenty-five-cent piece, while its feather-weight cousin, the pygmy shrew, *Microsorex hoyi,* may weigh little more than a dime. This makes it the smallest mammal on the continent and one of the smallest in the world.

Besides their size, shrews have another characteristic that makes them difficult to observe — their frenzied pace of life. In the summer, for example, masked shrews whisk through leaf litter, tunnel into moss, nip into rodent burrows, and scurry along streambeds, dashing here, darting there, all the while chittering noisily and quivering their long, flexible snouts. At night they are even more active than during the day, and in the winter they keep up the same hectic performance in runways under the snow. Everything about them is quick. They even sleep in a hurry, generally napping for about an hour at a stretch. In one minute, they breathe 850 times, and their tiny hearts pump out 800 pulses of blood. Under stress, their heart rate races up to 1,300 beats per minute, so fast that they may actually die of fright. A loud noise will sometimes scare them to death, and being captured by humans can send them into shock.

Interestingly, there is a connection between the shrews' high-strung, speeded-up natures and their diminutive size. Very small mammals have special problems with conserving body heat. This is the result of a basic rule of mathematics: the smaller the diameter of a cylinder, the higher the ratio between its volume and surface area. And what is a shrew, in its overall shape, except a tapering cylinder with ears and a tail? In practical terms, this means that shrews have relatively little body mass with which to produce heat and a great deal of skin from which it can radiate. They can't even insulate themselves properly because they are so short: a shrew in a thick coat would high-center on its fur, with its feet held up off the ground. Thus, shrews have no choice but to burn extra fuel to keep themselves warm. This accounts for their high rate of

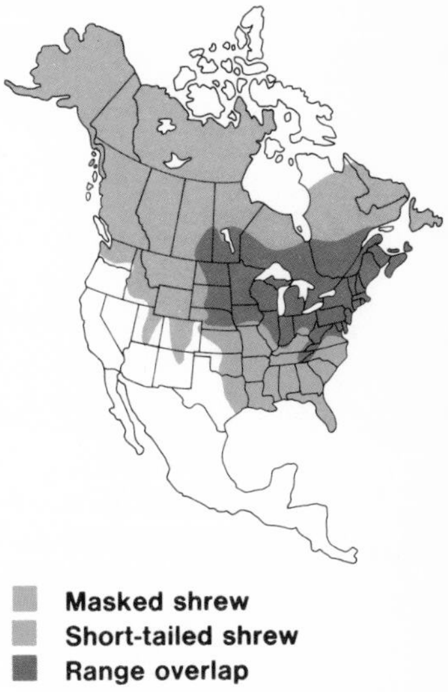

metabolism and the rush in which they live their lives.

The main business of the shrews' existence is getting enough to eat. Because of the constant chewing, their teeth wear down rapidly, with the result that shrews enter toothless old age after a year or eighteen months. In the wild, masked shrews commonly eat their own weight in food within twenty-four hours, and in captivity, they have been known to devour more than three times that much. Their favorite foods are fleshy insects, but they will also consume seeds, worms, mollusks, amphibians, bird nestlings, and mammals, including other shrews. The Inuit used to frighten themselves with the fancy that shrews could kill people by burrowing through to their hearts, though, happily, that isn't true. It is true, however, that short-tailed shrews can make short work of meadow voles and have been known to attack and kill animals the size of small rabbits.

If you stop to think about it, this is an incredible feat. How can any animal prey on something so much larger than itself? In the case of the short-tailed shrew, the answer is as simple as it is unexpected — poisonous saliva. People bitten by a short-tailed shrew complain of swelling and burning around the puncture and shooting pains that persist for a week. The effects on smaller animals are even more devastating, for they include convulsions, insensibility, and death. Though this "secret weapon" undoubtedly helps the short-tailed shrew to subdue voles and other mammals, it is more effective on such things as snails, earthworms, or beetles, which can be paralyzed with a shot of the poison and then cached in underground chambers. In this way, the short-tailed shrew keeps its larder full of preserved meat.

The ability to stockpile food is rare among insectivores. As a result, most shrews must forage constantly, summer and winter alike. One might expect many of them to starve during the cold months, when the insects disappear, but this does not appear to be the case. A researcher who studied two species of shrews in British Columbia discovered that, except for the few individuals taken by weasels, owls, and other small predators, almost all the shrews that headed into the winter survived to enjoy the following spring. Like most shrews, these species were territorial: each individual claimed exclusive use of enough habitat to get it through hard times. Thus, there was little competition over winter food supplies and no mass starvation.

Short-tailed shrews mark their territorial boundaries with feces and with the secretions of special scent-producing glands. In shrew society, these substances function as "no trespassing" signs. If a wandering shrew chooses to ignore the markers and trespass on a territory, the resident animal screams out its indignation. Often, this is enough to send the intruder into flight, with the ungracious host in loud pursuit. Should the interloper decline to leave, then the fight is on, with the two miniature combatants locked in a ball and rolling over one another on the ground. As a rule, these scuffles are won by the resident animal.

Most shrews are sociable only when they're very young. They can be born any time from spring to fall, since females sometimes have two or three litters in a season. (The number of offspring varies, though it is usually around six.) In the case of masked shrews, the infants' first home is a mass of soft plant fibers hidden under a log, stone, or stump, and here the wee newborns, naked and blind at first, huddle together in a ball. If they have to travel for some reason, they may form into a single-file "caravan," with each animal biting hold of its neighbor's rump and mother leading the way.

After a few weeks of nursery life, young shrews have generally had enough of society and may be ready to mark out territories of their own. Once that has been accomplished, they do their best to avoid one another until it's time for them to mate, when the males and females somehow manage to tolerate each other's company for a little while.

**Shrews have minute eyes (all but invisible in this photograph) and poor vision. Several species including the masked shrew, shown here, possess a crude form of echolocation, which helps them to detect obstacles but is probably not sensitive enough to assist in finding prey.**

## Order Chiroptera

# Bats

Primitive insectivores, the distant ancestors of today's shrews and moles, are thought to have branched into other evolutionary pathways, including one leading to the order of bats, the Chiroptera. There are hundreds of species of bats in the world, including such exotic families as the vampire bats of Latin America, which consume only blood, and the flying foxes of India and Madagascar, with wingspans of up to 1.7 meters. In western Canada there are about a dozen species of bats, all of them belonging to the family of smooth-faced bats (Vespertilionidae). The largest of these has a wingspan of only forty centimeters, and none of them consumes blood. Instead, like the ancient insectivores, they rely on a diet of insects.

To design a superb flyer from the flightless insectivores, natural selection conserved the basic mammalian blueprint but stretched it out of shape. In bats, the finger bones of the front limbs are thin and greatly elongated to form the delicate superstructure of the wings, while the thumbs are little more than claws, projecting forward from the wings at the wrist bones. A double-layered membrane of elastic skin stretches between the "fingers," back to the hind legs, and, in the smooth-faced bats, over the tail as well, forming a large but extremely lightweight flying surface. Even the hoary bat, *Lasiurus cinereus,* with its wingspan of forty centimeters, weighs only twenty-six grams.

Bats do their flying at night, and most of them expertly find their way without using their weak eyesight. During flight, a smooth-faced bat emits a series of very short sound pulses, which bounce off nearby obstacles in its flight path. These vocalizations are so loud that they must be nearly deafening to the bats, and yet we almost never hear them because most of the sounds are pitched well above the range of human hearing. Some of the reflected sound is scooped up by the bat's large, mobile ears and provides it with a radar-like picture of its environment. Unusual though it may be, "seeing with one's ears" is a skill not confined to the bats. Whales, dolphins, porpoises, some cave-nesting birds, and even some species of shrews use echolocation as well.

With their sophisticated sonar equipment and their great agility in flight, the smooth-faced bats are masters of an enormous feeding niche, that of night-flying insects. Bats are often seen silhouetted against a dusky, evening sky, fluttering in seemingly erratic flight while pursuing up to 150 insect meals per hour. For a typical species, such as the little brown bat, *Myotis lucifugus,* an aerial skirmish with an insect begins at a range of about one meter when the insect target is first detected by the bat's radar, which operates at a cruising rate of about 20 pulses per second. Suddenly the bat veers toward its target, and its echolocation pulse rate accelerates, peaking at a rate as high as 250 per second as the bat seeks more detailed information on the size, range, and speed of its quarry. If the chase is successful the insect may be seized with the teeth. More often, the insect is captured first with the bat's wing tip or with a temporary pouch formed from the tail membrane. The bat eats small insects in flight but may alight to pick the wings off larger ones before eating them.

Many moths (and some other insects) can hear ultrasonic sounds and take evasive action

**Like other tree bats, the hoary bat usually spends the day roosting in a shrub or tree. This one is holding its head upright, a sign that it is on the alert.**

Facing page: hanging upside down, torpid little brown bats blanket the ceiling of a cave.

when they detect an airborne attack by a bat. If they can't avoid the bat by flying away, certain species of moths may plummet to the ground, wings folded, or escape in a spiralling tailspin. Some tiger moths engage the bat in even more sophisticated warfare — they jam the bat's radar. In response to intense bat echolocation calls, these moths produce a burst of clicks with an ultrasonic component that accurately mimics the echoes from the bat's calls. The bat avoids the moth, possibly because it is confused by what it perceives as obstacles in the target's vicinity.

Cold and the absence of flying insects during the winter pose problems for bats at our latitude. According to the solution that they adopt, they fall into two groups: cave bats, which hibernate in caves or abandoned mine shafts; and tree bats, which migrate to the southern United States and beyond, where they remain active all winter or hibernate in hollow trees. The little brown bat, the species

The most common species of bat in western Canada is the little brown bat.

Little brown bat

with the widest distribution in Canada and the northern states, typically hibernates in humid caves where temperatures are just above freezing. With heartbeat and breathing rates slowed down, and their body temperature close to that of their environment, the bats slowly metabolize their winter reserves of fat. Seemingly spellbound, they may cluster on the ceilings in the dark stillness of the cave, hanging upside down by the sharp claws of their feet. Unless disturbed, they shake off their torpor only rarely, perhaps to drink from a subterranean pool or to lick droplets of moisture that have condensed on their fur.

With the arrival of spring, the little brown bats emerge from hibernation and disperse over large areas. The males tend to be solitary, roosting torpidly on the sides of buildings, under shingles, or beneath the bark of trees during the day. In early May, the females form maternity colonies of a few dozen to several thousand individuals. They are commonly found in attics or barns where the stifling heat of the sun promotes the rapid development of the embryos that have been growing since the females came out of hibernation. (Most matings take place just before hibernation begins, and the sperm is stored in the female's uterus throughout the winter.) The summer colonies are usually located near a pond or a stream, where the bats drink on the wing by skimming the surface of the water.

In June, each female little brown bat gives birth to a large baby. Born blind, pink, and naked, the newborn immediately clamps its teeth onto a nipple. If the mother is startled into flight, it clings there tenaciously, but at night, when the mother flies feeding sorties, the baby bat probably stays in the nursery.

When the mother returns, she locates her own offspring amidst the hordes of bickering bats by exchanging ultrasonic calls with it and confirming her identification by scent. By the end of July, when the young can fly, the bats leave the crowded nurseries.

Over the next few weeks, they migrate back to their traditional hibernation sites. On these migrations, they display uncanny powers of navigation, the exact nature of which remains a puzzle to biologists. Displaced from their home sites, little brown bats have been known to return over distances of up to 430 kilometers in only seventeen days. Blindfolded, they don't perform so well, but still may return home from a distance of 60 kilometers.

The low reproductive rate of little brown bats and a life span that may reach thirty years suggest that they have few predators. One threat to their existence is posed by people who explore hibernation caves in winter. Any disturbance at this time can cause the bats to arouse from their torpor, robbing them of some of the precious energy reserves that they need in order to survive until spring.

Disease is seldom a cause of death, but one of their afflictions is of concern to humans. Like all other mammals, they are susceptible to rabies, and about one out of every hundred little brown bats may be affected. For this reason, bats should never be handled with bare hands, particularly if they show obvious signs of sickness.

## Order Lagomorpha

# Pikas, Rabbits, and Hares

The rabbits and hares, which collectively form the leporid family, have few close relatives. The order to which they belong, Lagomorpha, contains only one other family, the pikas (Ochotonidae), which are distinguished by small, rounded ears, short limbs, the absence of an external tail, and some differences in dental and cranial characteristics.

Most of us, if asked whether or not the lagomorphs have any close living relatives, would think of those small, gnawing mammals, the rodents. Taxonomists used to be fooled, too, and formerly classified the lagomorphs as a suborder of rodents. Rodents and lagomorphs do, after all, share many common adaptations to a herbivorous mode of existence, including, for example, specialized dentition. In both orders, the incisor teeth, well-suited for gnawing and clipping vegetation, grow throughout life and are self-sharpening. There are no canine teeth, and a wide gap separates the incisors from the cheek teeth, which are adapted for grinding. It is now known, however, that these similarities are superficial, the result of different evolutionary stock developing similar ways of living (an example of convergent evolution). Closer examination reveals some fundamental differences. For example, while rodents have a single upper pair of incisors, lagomorphs have an accessory pair, small and peglike, located just behind the principal ones.

Another peculiar trait of lagomorphs, of less interest to the taxonomist because it is not unique to the order, is their habit of reingesting some fecal pellets. Two types of droppings are produced: the familiar dry, green, ellipsoidal pellets, and black, viscous ones which are promptly eaten. This practice, which probably provides protein, calories, and B vitamins that are otherwise not available, is loosely comparable to cud-chewing amongst the cloven-hoofed mammals. Both processes aid in the tough task of digesting plant matter.

## Pikas

One of the special sounds that soon becomes familiar to anyone who visits alpine areas in the Rocky Mountains is a peculiar, piercing note that might be likened to someone crying "peek" in a high-pitched voice. This is the call of the Rocky Mountain pika, *Ochotona princeps,* a diminutive lagomorph that usually makes its home on the border between grassy meadows and rockslides. It is generally found near the timberline in the mountains but also lives near sea level on the coast of British Columbia. A pika is unwary of human intruders, almost to the point of appearing foolhardy in late summer and fall. Yet it is much easier to hear than to see. About the size of a hamster, its round, nearly tailless form

easily blends into its boulder-strewn background, especially since each regional population of "rock rabbits" is colored to match the local rocks. Besides its frequent call, there are other signs of a pika's presence, such as piles of distinctive, small, round droppings and conspicuous white stains on a few rocks where it habitually urinates.

Nearby, in an open area or sometimes under a boulder, a miniature haystack may be found, up to half a meter in height and as much as a meter wide. This mass of dried and drying vegetation is a major element in the pika's strategy for surviving the long, alpine winter and is the focal point for its activities throughout the year. Instead of hibernating like the ground squirrels and marmots that share the rocky slopes, or fronting the elements like the hardy mountain goat, the pika shifts its activities to the environment beneath the snow where it can munch on its haystack, graze on low-lying plants such as saxifrage and moss campion, or gnaw the bark from the base of aspen trees.

The construction of the haystack is an individual project for each pika, begun as early as June and invariably well under way by August. The pika literally makes hay when the sun shines; unless the weather is wet or especially hot it may go on more than a hundred foraging trips each day, traveling along well-worn trails that lead from its rocky retreat to the nearby meadows. Each of these trips takes about a minute and nets another mouthful of vegetation to be deposited on the haystack or, occasionally, onto two or more smaller stacks situated beside one another. Many of the plants that grow in the vicinity may show up in the pika's mound of goodies: hedysarum, gooseberry, willow, buffalo berry, lupine, vetch, dwarf huckleberry, evergreen sprigs, grasses, and sedges, for example. Droppings of marmots and coyotes are also stored and consumed, presumably to satisfy some nutritional requirement. Inedible materials in the food pile may include an underlying layer of leftover twigs and straw if, as is common, the pile has been built on a previously used site;

Facing page: the collared pika, *Ochotona collaris*, wears an indistinct light gray band around its neck. It is thought to be a separate species from the more southerly Rocky Mountain pika.

In windy areas, a Rocky Mountain pika often chooses a sheltered spot for constructing its haystack.

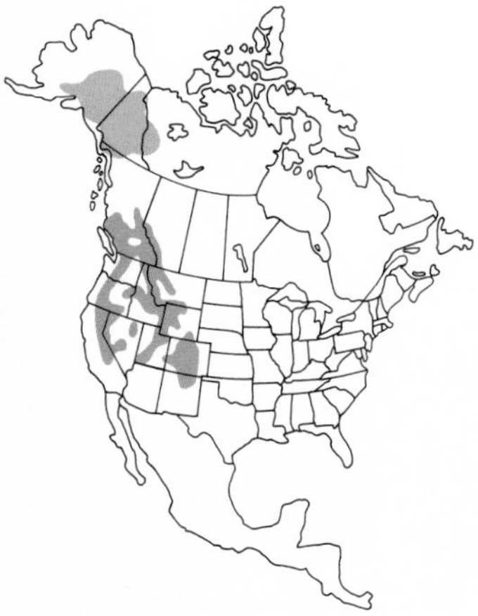

Collared pika
Rocky Mountain pika

and a topping of sticks, stones, and other debris, added in late September. Such a roof may protect the haystack from wind and weather, though most likely it is added simply to satisfy the pika's acquisitive urges when fresh vegetation is no longer available.

With so large an investment of time and energy at stake, pikas are understandably possessive about these hoards. Although they regard the nearby meadows as a sort of community pasture, they lay exclusive claim to a rocky area, a few hundred square meters in extent, surrounding the haystack. Their aggressiveness towards other pikas, coupled with their need to stay within about fifty meters of their foraging grounds, means that, on large rock slides, pika territories are regularly spaced around the perimeter of the rocky area. Maintenance of these territories is accomplished mainly by chasing off trespassers, though sometimes there are fights, generally harmless skirmishes involving a few kicks with the hind feet, a few bites, or a boxing match with the forepaws. Owners advertise their presence in a territory by scent markings (with urine or by rubbing a cheek gland against rocks) and by keeping up a steady exchange of vocalizations with their neighbors all day long and well into evening.

The "peek" call is used in this context and consists of one or, sometimes, two short notes, often repeated many times at intervals of a few seconds. If a predator such as a marten, coyote, hawk, or owl appears, this call is used to raise the alarm. Pikas, hoary marmots, and Columbian ground squirrels sometimes respond to each other's warning calls by interrupting their activities to locate the source of the disturbance. But the normally vocal pikas are significantly silent when the deadliest of their enemies, the ermine and the long-tailed weasel, put in their appearances. These slender killers can easily follow the pika into any burrow or crevice, so the pika's best defense is to remain inconspicuous and not draw attention to itself.

The pika has another call, a chatter followed by two to six "peek" notes, which is heard most often at the height of the breeding season in May and June. All yearling and adult females are potentially capable of bearing one or, more often, two litters by the end of the summer, the second pregnancy following immediately after the first. Litters of two to four young are born in underground dens after a gestation period of thirty days. Childhood is short for pikas: when only a few weeks old they appear above ground, already weaned and ready to commence the adult activities of foraging and building haystacks. And within a week of their emergence they are driven out of their home territory by the mother and left to fend for themselves.

# Cottontails

**Although Nuttall's cottontail, (shown opposite) is paler than its near-relative, the eastern cottontail, the two species can most easily be distinguished on the basis of their distributions. Both are capable of reproductive prodigies.**

The western Canadian cottontails are the most northerly members of a farflung southern tribe. In contrast to their snow-adapted relatives, the hares (most of which don camouflage coats of white for wintertime), cottontails retain brown coats the year around. But in spite of this handicap, the cottontails have been able to enlarge their western Canadian range in recent times. The eastern cottontail, *Sylvilagus floridanus,* didn't occur here until about 1910, when the species started moving north through the Dakotas. The slightly smaller Nuttall's cottontail, *Sylvilagus nuttallii,* which is native to the shrubby bottomlands of southeastern Alberta and southwestern Saskatchewan, has also been increasing its range to the north and east. Surprisingly, these expansions are thought to be the result of farming, which, through the suppression of prairie fires and the planting of

shelterbelts, has provided the cottontails with the wooded cover they need for daytime hideaways and winter food. All the same, it must be a marginal existence: four cottontails in every five have sometimes been known to die in a wintry month.

Fortunately, cottontails are insured against extinction by their legendary ability to reproduce. In parts of the United States, female cottontails can produce seven or eight litters a year, and even in Canada, where the breeding season gets off to a later start, four litters of five or six bunnies each may be typical. That's a couple of dozen offspring per female every year!

If the cottontails' reproductive rate is remarkable, their mating behavior is nothing short of bizarre. By late winter, when the courtship begins, the male eastern cottontail is already in breeding condition, so he makes the first advances, with his ears up and head held high, looking eager and alert. The female, who has not yet come into heat, responds grumpily, folding her ears back flat against her head and crouching as if ready to pounce. If the male gets too close, she boxes at him with her front paws or chases him off. Sometimes, though, instead of retreating meekly, the male reacts by dashing past the female and urinating on her as he speeds by. This is apparently devastatingly seductive to a female cottontail, and, after as many as twenty repetitions, it may put her in the mood for the next step in the ritual. Instead of running past the female, the male rushes straight at her, causing her to leap up out of the way so he can scoot underneath. At the height of her jump, the female often urinates. The pair then face one another and go through the performance again, sometimes changing roles: the male may be the one to jump, clearing the way for the female to zip beneath. Few people ever witness these performances because cottontails are largely nocturnal.

Such displays of rabbit acrobatics play no part in the actual mating. Some biologists suggest that they simply serve as a way for the males to let off steam before the females come into estrus. When breeding does begin, it takes

place at about the same time throughout a given population, so that all the litters in a particular locality are born within a day or two of each other. Since remating takes place immediately after parturition, subsequent litters are also born in synchrony, about a month apart. This reduces the number of newborn bunnies lost to predators.

The eastern cottontails' eccentricities are not limited to their courtship habits, for they also have an unusual method of giving birth. The expectant female begins her preparations by scooping out a shallow burrow (about ten centimeters deep) and lining it with dry grass. Meanwhile, one or more males wait nearby, nosing in at every opportunity to sniff the female or inspect the nest. The doe tolerates this pestering through the early stages of nest-building, but by the time she is ready to add the finishing touch — a cozy padding of her own fur — she is out of patience with the intrusions. She may even attack by pinning a bothersome male on the ground and biting him so hard that he squeals.

The female can be forgiven for her bad temper, because she is in labor at this point. The actual delivery occurs with the female sitting over the mouth of the burrow. Just as the first of the bunnies are about to be born, the mother spins around several times. After a brief pause, during which some or all of the young apparently are born into the burrow, the female may whirl again. Another short pause, perhaps for nursing this time, and the female is ready to cover the nest and leave for several hours. (The young are only fed once or twice a day.)

Before the female can settle down to rest or browse, she has to deal with her male attendants, who take off after her at full tilt as soon as she leaves the nest. Within seconds (minutes at the most) the successful suitor will have mated her, but the other males continue to thrash about, colliding with one another and scrapping until long after the prize has been lost. Much of the mating is done by the one or two males that have established themselves at the top of the social ladder in their community. These are the animals best able to rout their

■ **Eastern cottontail**
■ **Nuttall's cottontail**
■ **Range overlap**

fellows in day-to-day chases and fights. Because they can intimidate or repel their rivals, they have a better-than-average chance of being the first to reach the female at mating time. There is evidence of a similar hierarchy among the females, which also has a bearing on reproductive success. In both cases, the dominance order helps to ensure that the superior animals produce the most young.

Cottontails need whatever evolutionary advantage this mechanism provides, for their lot is a difficult one. Almost any meat-eating animal or bird relishes a good meal of rabbit, and the list of their predators is long: snakes, crows, hawks, owls, weasels, foxes, dogs, people, and others. Though it can hardly come as any consolation to the cottontails, they and the hares are of prime ecological importance as converters of grasses, leaves, and twigs into food for carnivores.

# Hares

The hares are represented in western Canada by three species: the arctic hare, *Lepus arcticus,* a robust tundra-dweller; the snowshoe hare, *Lepus americanus,* which is equally at home on the plains or in the boreal forests; and the white-tailed jack rabbit, *Lepus townsendii,* a large plains-dweller which, despite its name, is a hare, not a rabbit. One difference between the two groups is that hares, unlike rabbits, are born fully furred, with their eyes open, and ready to hop about a few minutes after birth. They differ in appearance too: hares are generally larger than cottontail rabbits and have longer ears, the tips of which are black year round. And unlike cottontails, most hares change their coat color with the seasons to blend in with their background.

Of these three species of hares, the snowshoe is most widely distributed and is perhaps the most familiar, having achieved fame for its spectacular cyclic fluctuations in numbers. (Jack rabbits and arctic hares have similar

**Arctic hares often herd together by the hundreds. Their summer coloration depends on latitude — this group's only concession to the brief warm season is a bit of brown fur on their faces.**

**The snowshoe hare is also known as the varying hare because of its seasonal changes in coat color. This individual has nearly completed the spring moult.**

variations in population, but they have not been so closely studied.) The snowshoe hare cycle is synchronized over large regions and recurs every eight to eleven years. During peak winters, forests and thickets conceal an occupying army of snowshoe hares. In the fall of 1970, for example, the population in central Alberta reached 1,100 to 2,300 per square kilometer. At such times, their tracks are everywhere, and nearly every bush or low-lying tree shows where the nibbling multitudes have appropriated winter rations of bark or twigs. Their ranks swell the most in favorable habitat such as the margins of bogs and clearings or recently burned forests where there is lush growth of young woody plants. But a few years after the peak, the troops are mysteriously decimated, and the survivors are typically reduced to fewer than a twentieth of their former numbers.

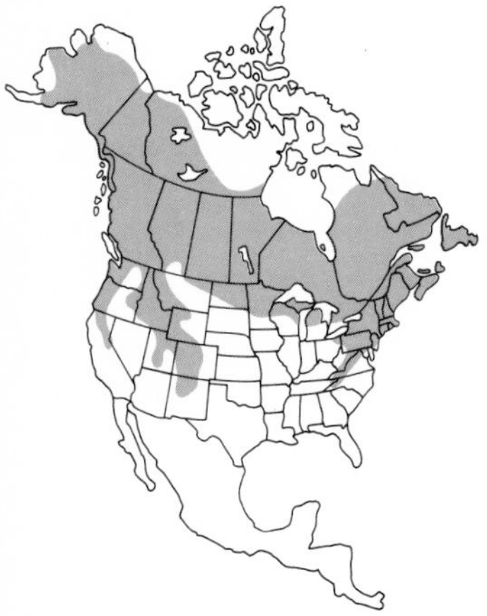

**Snowshoe hare**

What foe brings about these population crashes? The current view is that snowshoe hares are their own worst enemies. In summer, their taste for herbaceous plants is readily satisfied, but in the winter they must turn to bark and buds, reaching ever higher as the winter snow accumulates. During the years of increasing population, an often-conspicuous browse line is established, below which new woody growth is retarded. So while their numbers are increasing, their winter food supply is steadily decreasing. Eventually, a time of reckoning arrives. For two or three years, winter survival rates drop, litters are smaller, and the breeding season shorter. Predation becomes increasingly significant, further depressing the population and prolonging the period of decline. The shortage of hares then leads to a decline in the number of predators. Soon the remaining hares, many of which have survived in dense thickets where predation is difficult, begin to expand in numbers once again and spill over into less favorable habitats.

The periodic collapses and explosions in hare numbers reverberate throughout the community of carnivores that depend heavily on snowshoe hares for food. Lynx populations, for example, follow the ten-year snowshoe hare cycle and peak one or two years after the hares do. In times of snowshoe hare scarcity, lynx numbers drop to fewer than one-quarter of

The long, antenna-like ears and large body size of the white-tailed jack rabbit, shown here, help in distinguishing it from the snowshoe hare.

their former levels. The great horned owl, too, echoes this story of feast or famine. The first spring after a snowshoe hare peak, virtually every breeding pair of these owls is nesting, but four years later, as few as one pair in twenty may raise a brood. Many other predators, including the coyote, red fox, and goshawk, may also follow the snowshoe hare cycle.

The explosion of snowshoe hare numbers that leads up to the peak years would not be possible without the hares' enormous reproductive potential. They breed most prolifically in the prairie provinces, where each female over six months of age is capable of producing between eight and eighteen young every summer. To achieve this feat, snowshoe hares begin breeding by the first of April or sooner and bear their litters five weeks later. Not wasting any time, the females remate within a few hours of parturition and carry on to produce as many as four litters in a single season (fewer during a population decline). Successive litters are born at about the same time throughout a local population.

Snowshoe hares indulge in a variety of antics at the time of mating. The females may lead the males on a wild romp through the woods or shun their escorts by facing them, drumming the ground with their hind feet, or leaping up in the air. The male role in these performances includes urinating on the females and jumping. Amidst squawks and drumming, the males often engage each other in harmless battles, the object of which seems to be to kick one's opponent while leaping over him.

Outside the mating seasons, snowshoe hares are somewhat more sedate in their habits. They usually spend the day in secluded resting spots called "forms" and venture forth only in the evening. Midnight may find them engaged in nibbling back the vegetation that encroaches on their "runways" — elaborate trails that crisscross their home ranges (about eight hectares each), providing the hares and many other creatures with highways and escape routes from predators. Other common activities include stretching and rolling themselves in favorite dust wallows during summer, as a remedy against ticks; or, in the winter, if the snow is deep and loose, stomping out runways with the long, wide, hind feet that give the species its name.

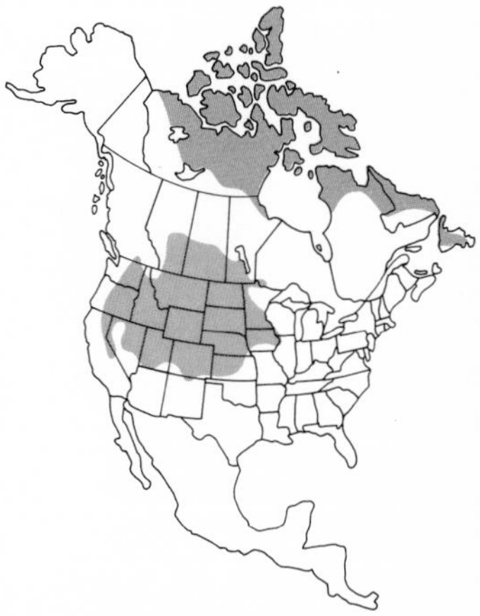

Order Rodentia

# Rodents

We're used to thinking of the primates, and *Homo sapiens* specifically, as the most successful of mammals, but in certain respects our accomplishments are paltry compared to those of the rodents, order Rodentia. Not only are they the most numerous of mammals, but they flourish in an overwhelming diversity of forms: about forty percent of all modern mammalian species are rodents.

To what do they owe their phenomenal success? In a word, to their teeth. Their most outstanding characteristics, and the ones that unite them taxonomically, are the gnawing and grinding capabilities of their dental equipment. Two prominent pairs of incisors, chisel-like in form and function, are the basic gnawing apparatus, which can be put to such diverse uses as felling trees, in the case of beavers, and dissecting seed heads, in the case of mice. These remarkable tools are designed to be self-sharpening and ever-growing: the front surface of each incisor is composed of hard enamel, the rest of softer dentine, so the abrasion of the teeth against one another leaves a sharp edge; to compensate for continued wear, the incisors (and the rest of the teeth, in some rodents) grow throughout life. If the incisors fail to wear fast enough, due to deformity, disease, or too soft a diet, the teeth grow in large arcs and may eventually pierce the throat and snout of the afflicted animal.

Since rodents are generally herbivorous, their teeth must also function to grind fibrous plant matter into a digestible pulp. This is accomplished by the molars, which are well developed and ridged with enamel to increase their efficiency. All herbivores must also possess modifications of the alimentary system to facilitate the digestion of cellulose. In rodents, an enlarged caecum (an organ that corresponds anatomically to the human appendix) serves as a fermentation vat where microorganisms aid in processing the food. After the first passage through the gut, via the caecum, the droppings are soft and contain abundant vitamins. This accounts for the reingestion of feces that is recorded in many species of rodents: digestion isn't complete until a subsequent passage of the same material through the gut produces dry, hard droppings.

Incisors for gnawing, molars for grinding — this completes a rodent's need for teeth. The rest, including second incisors, canines, and front premolars, have been lost in the course of evolution. The resulting gap, or "diastema," is more useful than the missing teeth would be, for by drawing the cheeks together in the diastema, a rodent can close off the back of its mouth, thereby preventing debris from entering while the incisors are at work.

Another refinement of the rodent mouth is its dual-action jaw. When the molars are being used, the lower incisors lie behind the upper pair. With the lower jaw brought forward, the incisors are engaged, but the molars do not meet and consequently do not wear against one another. The complex muscles and skull bones that are associated with these specialized jaw movements are used by taxonomists as an aid in classifying the thirty-four families of rodents that occur in the modern world.

The astonishing variety that is characteristic of these families is well displayed in the eight that are native to western Canada. One or more families are represented in almost all of our life zones, from arid plains to coastal rain forests to frigid tundra. In the trees, on the ground, in the water, or beneath the surface of the earth, they make their livings by running, hopping, swimming, burrowing, even gliding through their chosen environment.

One of the better burrowers is the mountain beaver, *Aplodontia rufa* (the only member of the family Aplodontidae), which inhabits southern British Columbia and the American Northwest. Its compact body, short legs and ears, small eyes, and strong claws are all adaptations to its fossorial habits. It isn't highly specialized though and may also climb trees and swim with ease. Its lack of specialization is apparent, too, in the anatomy of its chewing muscles and related bones, which reveal that the mountain beaver is by far the most primitive of the present-day rodents, a unique, living fossil. More extreme in their adaptations for burrowing are the pocket gophers (family Geomyidae), which lead primarily subterranean lives. They are found in mountain valleys of south-central British Columbia, on the Canadian prairies, and throughout most of the United States. All that one usually sees of them are the results of their excavations — entrance-less mounds of fresh earth (often erroneously attributed to moles) and "gopher cores," which are rope-like patterns of earth that appear, after snow-melt, in areas where the animals have pushed soil into snow tunnels.

Closely related to the pocket gophers are the pocket mice and kangaroo rats (family Heteromyidae) which, like the "gophers," have external, fur-lined cheek pockets that are used for carrying seeds and other food items. Members of this family are mainly plains dwellers and possess many adaptations for life in arid climes, including, for example, their ability to subsist for many weeks without water. (They extract all of the moisture that they require from the insects and dry seeds in their diet.) Unlike the pocket gophers, the pocket mice and kangaroo rats are hoppers, using their long hind legs for propulsion and their long tails for balance. Another family of hoppers includes the jumping mice (family Zapopidae), which are widely distributed throughout the prairies and boreal forest. At the approach of a predator, some of these diminutive creatures can spring off in bounding leaps of up to two meters.

Better known than the burrowers and the leapers are the families that form the subject of this chapter. The squirrels (family Sciuridae) are a large and varied assemblage, ranging from the terrestrial chipmunks, woodchucks, marmots, and ground squirrels to the arboreal red squirrel. Some of their numbers, the flying squirrels, are capable of gliding, making them the only mammals other than the bats to have taken to the air. The rodent line has evolved some semiaquatic forms as well, including the beaver (family Castoridae) and the muskrat, which belongs to the family of rats, mice, and voles (Cricetidae). Also included in this chapter is one of nature's most bizarre experiments, the porcupine (family Erethizontidae).

# Chipmunks

The chipmunk is an alert and energetic personality, readily encountered in the wild. Often this diminutive member of the squirrel family introduces itself with a burst of high-pitched chatter and a rustling of leaves as it scampers through the litter on the forest floor, in a performance designed to startle and confuse a potential predator. At other times, one hears its steady "chip, chip, chip . . ." as it signals to other chipmunks, and to many other mammals and birds in the woodland community, that a potentially dangerous intruder is approaching. For all its verbal protesting, it doesn't seem very wary though and may easily be observed

The yellow-pine chipmunk, shown here, can be distinguished from the least chipmunk by the black coloration on the fronts of the ears. On the least chipmunk, by contrast, the fronts of the ears are tawny.

as it dashes along a fallen log, scurries up a tree trunk, or pauses briefly on a rock to survey its surroundings.

There are five very similar species of chipmunks in western Canada. The two that are most widely distributed, the yellow-pine chipmunk, *Eutamias amoenus,* and the least chipmunk, *Eutamias minimus,* can both thrive in the same wide variety of habitats, from near desert, to forest, to alpine meadow. Their environmental niches are so similar that in some areas where the ranges of these two species overlap, such as western Alberta, they cannot actually coexist. Here "competitive exclusion" limits their distributions to different altitudes: sunny, forested areas are the domain of the yellow-pine chipmunk, which easily dominates the slightly smaller least chipmunk in aggressive encounters; but talus slopes and the area above the treeline are claimed by the least chipmunk — its smaller size (about forty grams versus about fifty-five grams) and a slightly lower birthrate give it an advantage in these areas where food is scarcer.

The diet of chipmunks is typified by that of the least chipmunk and includes grains, sedges, fruits and their seeds, nuts, roots, bulbs, and insects. In the summer and especially in the fall, the least chipmunk spends much of each day at the task of gathering these varied and scattered bits of food. A great aid in its labors is the dexterity of its front paws, which are as useful for pulling down grass stems, hand over hand, to reach the seed-laden heads, as they are for manipulating strawberries or gooseberries to remove the seeds. Food that isn't immediately eaten is crammed into special pouches on the insides of the cheeks and carried off to be hidden in an underground burrow or in small, shallow pits dug close to where the food was gathered. Many of the small caches are destined later to be unearthed and eaten or added to the main underground store, but many seeds are missed and eventually germinate. In this way the chipmunk contributes to the reproduction of those plants that it depends upon for its livelihood, making some compensation for the large numbers of seeds that it eats.

The least chipmunk's burrow is usually excavated in a wooded area where, in winter, the protection from wind will allow snow to build up a deep, insulating blanket. Usually, a well-concealed entrance leads to a single, gently sloping tunnel, about a meter in length. At the end of the tunnel is a nest chamber, about seventeen centimeters in diameter, lined with such materials as rabbit fur, poplar or willow cotton, deer hair, shredded feathers, and plant fibers. The least chipmunk sleeps in its pantry: its food hoard, consisting of up to a kilogram of seeds, is cached underneath and beside the nest.

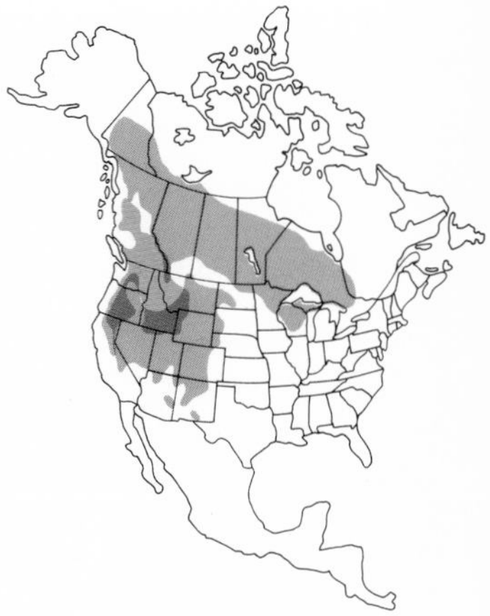


As fall progresses into winter, the least chipmunk spends more and more of its time ensconced in this cozy, well-stocked hideaway, appearing outside only during the warmest part of the day. By November in southern Canada, or early October in the northernmost part of its range, it is spending most of its time in hibernation. Compared to other hibernators like bats and marmots, the least chipmunk's "sleep" is fitful. Its body temperature doesn't drop very low, only to ten degrees Celsius, and every four days it rouses itself for several hours, bringing its body temperature back up to normal. Chipmunks store very little body fat for the winter; instead, they use these wakeful periods to nibble away at their provisions. On unusually warm days, they may even go out.

The males are the first to emerge from hibernation, sometime in April. A couple of weeks later, the females appear, and soon, in the midst of many chases and fights, breeding takes place. Litters of four to seven young are born near the end of May and are raised by the mothers, often in underground burrows. Litters may also be raised "outdoors," in nests reserved for summer use only. Such a nest may be in a hollow tree or log, in an abandoned woodpecker hole, or may simply consist of a hollow ball of grasses, resembling a bird's nest, lodged high in a conifer tree. The young are weaned at five or six weeks of age and disperse in August or September to establish burrows of their own in preparation for the coming winter. They breed in their first or second year and, with luck, may live to reach six years of age.

# Woodchuck

One of the few species of mammals to have benefited from the introduction of agriculture to the prairies is the woodchuck, *Marmota monax*. Over much of its range it is still a forest dweller, but it also thrives in cultivated areas, where it finds garden vegetables and alfalfa or corn crops tasty additions to its natural diet of wild grasses and leaves.

In the woodchuck, nature seems to have perfected the digging machine. In light soil the "groundhog" can bury itself from sight in a minute, with its strong front claws tearing into the ground and its hind feet ejecting a spray of earth backward. Using its teeth, it chisels out stones or clips off roots, and with its forepaws and flat head, it bulldozes loose soil or stones from its tunnels.

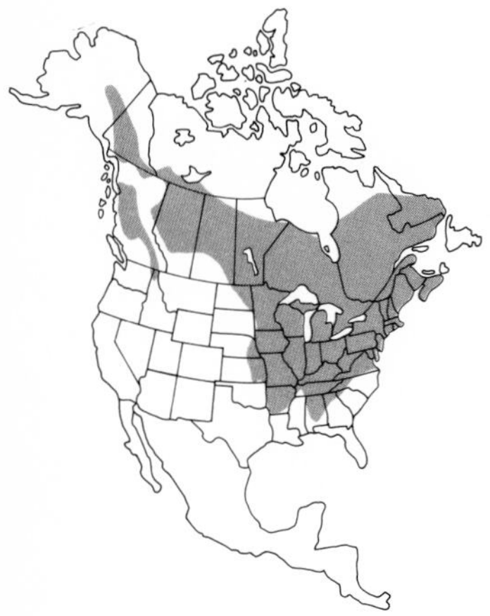

Woodchuck

While woodchuck burrows vary greatly in their complexity, a typical dwelling incorporates some sound engineering practices. From the main entrance, which is indicated by a large mound of fresh earth, a tunnel leads to a chamber about a meter beneath the surface of the ground and then forms several branches that slope upward, to prevent flooding. One of these branches terminates in a nest chamber lined with grasses or shredded leaves; another may lead to a toilet chamber where excrement is deposited. By burrowing upward from one of the tunnels, the woodchuck provides itself with another entrance, lacking a conspicuous earth mound. This "plunge hole" drops straight down for half a meter or so, providing instant refuge for a woodchuck fleeing from a predator.

For summer use, the den is typically located in flat or rolling countryside, in an area with good drainage and ready access to a favorite feeding area. For winter hibernation, another simpler burrow is sometimes excavated, preferably on a slope where drainage is good and in a forest where the nest chamber can be dug under the roots of a tree, giving maximum protection against predators.

About a month before hibernation begins, a woodchuck is already occupying its winter quarters, where it resides alone or occasionally with a single companion. By this time the woodchuck is extremely plump, about forty percent of its body weight being fat. In late September or early October, it barricades itself into the hibernation chamber, packing the soil behind it. Then the round 'chuck rolls into a tight ball and enters a deep torpor, with its body temperature and pulse rate lowered and its breathing slowed to once every six minutes. Its store of fat will last it until breakfast, more than six months later.

Like restless sleepers, woodchucks rouse themselves at regular intervals during the winter and may even wander about if the weather is mild. Around March, woodchucks begin to emerge for the summer season, sometimes digging through a crust of snow to end their entombment. But their fasting continues and they must rely on their remaining fat reserves for several weeks, until new greenery appears. The sexually mature males (those two years of age and older) are out and about first, approximately three weeks before the yearlings and females. Within a month of their emergence, the males have mated with as many females as possible, and both sexes are in their summer dens, living a solitary existence. Then the only common interaction amongst adult woodchucks is mutual hostility. When neighbors meet, they may threaten each other with bared teeth and arched backs or even engage in a fight.

In late May or June, depending on the climate, litters of about four young are born, blind, almost naked, and completely helpless. By the time they see the full light of day, about a month later, they are almost weaned and their mother has introduced them to green food. At this stage, when an infant asks to nurse by squeaking at its mother, she may disdain its request by gently placing a paw on its head. A persistent youngster may succeed, or it may be rebuffed by an aggressive chatter of teeth or a slap with a paw. About a week after their first emergence, the young 'chucks seem to tire of this treatment and disperse to dig their own dens or, more likely, to appropriate abandoned ones.

For the rest of the summer, juveniles and adults alike take life easy. In spring and fall they prefer to be above ground only at midday, when it's warmest, and even in midsummer, when they are active morning and evening, their total time spent out-of-doors is only about three hours per day. If their dens are not situated in the midst of good forage, they may commute along well-worn runways to favorite pastures or meadows, seldom more than thirty meters or so from their dens. Constantly on the alert for danger, they frequently rise on their haunches to look about, ever ready to dash down a hole or give a warning with a shrill whistle that has earned them the name of "whistle-pig." They enjoy basking in the sun, usually on top of their mounds, but sometimes they surprise a human observer by loafing high up on the limb of a tree!

In areas where woodchucks are abundant, they play an important role in the natural scheme of things. Their dens are used for family homes or temporary shelters by a long list of species, including skunks, raccoons, river otters, and occasionally even chipmunks, mice, and voles. Many a hibernating woodchuck has been the unwitting host for a cottontail rabbit seeking shelter from a cold winter wind. And the red fox readily usurps a woodchuck's lodgings as well and may go on to add the woodchuck to its menu.

The red fox is the main predator of woodchucks, especially juveniles, but it is sometimes evenly matched by a full-grown woodchuck, both in weight and ferocity. Perhaps this helps to explain an unusual and rare phenomenon: a woodchuck and its young sharing a den with its arch enemy, the red fox.

**One of the larger members of the squirrel family is the woodchuck. The average adult weighs about three kilograms, but some are twice that size.**

# Marmots

The woodchuck has three close relatives in western Canada, all of which belong to the same genus, *Marmota*. One of these is the yellow-bellied marmot, *Marmota flaviventris*, or "rockchuck," so called because it makes its home in rocky meadows or talus slopes. The Vancouver Island marmot, *Marmota vancouverensis*, is found only on a few mountains of Vancouver Island, near the treeline. It is very rare, the total population probably numbering fewer than a hundred. The most common of our marmots is the hoary marmot, *Marmota caligata*, an animal that is easily noticed by anyone who ventures into its alpine and subalpine habitat. Long before it is seen, it usually sounds a warning with a high-pitched whistle.

Although they closely resemble woodchucks, hoary marmots are surprisingly different in their social lives. Whereas woodchucks prefer solitude for much of the year, hoary marmots live in colonies that typically include several adults and a dozen of their immature offspring. Each colony provides itself with abundant accommodation by tunneling under large boulders where predators such as the grizzly bear cannot easily dig its members out. Shallow dens, used for an overnight stay or temporary shelter, seem to be viewed as communal property. But a deeper burrow may be used by a female with young under a year of age, in which case she tolerates no other company, not even the dominant adult male. He either lives alone or moves in with an adult female and her yearlings.

One of the advantages of a communal existence is that, collectively, the animals are more alert for possible predators than any single individual could be. When they aren't feeding, hoary marmots often lie on rocks or on the mounds of earth at the burrow entrances, stretching or grooming while keeping a sharp eye out for intruders. If one of them sees trouble coming, it not only sounds the alarm, it specifies the nature of the disturbance by the type of call given. For example, the long whistle that often greets a human intruder may mean something like "possible danger," because it signals other hoary marmots to stop what they are doing and look about. A descending call, often given if a bird of prey suddenly appears, carries a more urgent message, such as "Run for your lives!" for it sends the whole colony scampering to their burrow entrances where they can quickly plunge from sight if need be. Besides their vocal repertoire, which includes other calls, chirps, and growls, they seem to use long-distance visual communication as well. When going anywhere, all but the very youngest give a quick flip of the tail before moving, perhaps to avoid causing alarm amongst their fellows by being mistaken for a predator.

In other ways as well, hoary marmots show their companionable natures. They frequently meet one another on trails leading to burrows or feeding areas and engage in enthusiastic greetings. Initially they touch their noses or mouths together, probably to exchange identifying scents. Then they may chew on one another's faces, necks, and backs. If they are feeling especially playful, one of them may throw down the gauntlet by cuffing the other with a paw. Then the match is on! Good-naturedly, they roll and wrestle together for minutes on end, biting at the fur on each other's throat and head, and boxing with their forepaws. Sometimes these antics develop into pushing matches with both animals on their hind legs, noses pointed skyward, as each strains to topple the other over backwards.

In the springtime, there is a serious element to these encounters as well. If an adult is involved, it will likely attempt to rout its opponent. Most often, it is the two-year-olds that are harassed in this way. This is because the young of the season don't emerge until late July, by which time chases are infrequent. The yearlings are seldom affected because they are still tolerated by their mothers in the home

**At an average adult weight of six kilograms, the hoary marmot is the largest member of the squirrel family. Nicknamed "the whistler," its shrill alarm calls are heeded by pikas, ground squirrels, and other marmots.**

burrows. But when the young are two years old, their mother gives birth again and turns out the old litter to fend for themselves.

By having one litter every two years, hoary marmots set a record for slowness of reproduction amongst rodents. Very likely, this is a consequence of the short growing season at high altitudes. Within the brief span of about three months, the mother must consume enough fresh green plants to enable her to bear and nurse a litter, then accumulate fat for the winter hibernation. So demanding is this task that a mother hoary marmot, with her infants, enters hibernation a couple of weeks later than other colony members, because she needs to forage as late in the season as possible. She doesn't really catch up on her eating until the next summer, so she isn't ready for another pregnancy until the summer after that. The short growing season may also explain why hoary marmots are so much more sociable than their lowland cousins, the woodchucks. Living in areas with long growing seasons, young woodchucks put on weight fast, enabling them to disperse in the season of their birth and to reproduce as yearlings. Young hoary marmots, on the other hand, grow more slowly and don't achieve adult size and sexual maturity until they are three years old, so this species has evolved a social order that provides protection and guidance until the young are two years old and ready for the rigors of dispersal.

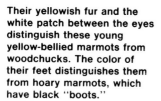

**Their yellowish fur and the white patch between the eyes distinguish these young yellow-bellied marmots from woodchucks. The color of their feet distinguishes them from hoary marmots, which have black "boots."**

# Ground Squirrels

Everyone can recognize ground squirrels, whether by their size and shape, the flick of their tails, or the chatter and squeak of their calls. On the prairies in particular, they sometimes seem to be everywhere, drawn pompously up onto their haunches like miniature sentries. Yet for all their familiarity, few of us can claim to know them very well. There are seven species of ground squirrels in western Canada: the arctic, Columbian, golden-mantled, Cascade golden-mantled, Franklin's, Richardson's, and thirteen-lined, by name. Each is distinct in appearance, distribution, and patterns of social behavior.

The most familiar member of the group is Richardson's ground squirrel, *Spermophilus richardsonii,* better known in everyday parlance and folklore as the prairie "gopher." This common name (which, unfortunately, has also been bestowed on four other species of ground squirrels, another genus of burrowing rodents, a tortoise, and a snake!) derives from the French *gaufre,* or honeycomb, a reference to that part of the squirrel's domain which we never see — its maze of underground passageways. Every adult Richardson's ground squirrel has one or more burrows to call home. Some are simple tunnels, less than four meters long, but an elaborate set of digs will wind along for about fourteen meters, issuing now into a "grain bin," now into a toilet compartment, now into a dead end. The softly padded nest chamber is in the deepest part of the labyrinth. Here the squirrels sleep, hibernate, and rear their young. All told, they spend ninety percent of their lives underground.

The first Richardson's ground squirrel to emerge in the spring usually pokes its head out in March, though the exact time varies with latitude, altitude, and weather. These early risers are invariably males overcome with wanderlust; they can't even wait for the soil to start thawing before they're on the move. Female Richardson's ground squirrels, in contrast, are homebodies: they usually take up residence near their mothers and occupy pretty much the same area year after year. If it weren't for the springtime travels of the males, the population would become inbred, and it might be "under-bred" as well. Although equal numbers of males and females are born, adult females outnumber the males by as much as four to one. If all or nearly all the females are to reproduce (which is the general rule), at least some of the males must scurry around and mate several times each spring. This happens immediately after the females come out of hibernation, a week or two later than the males. Once the mating is taken care of, the males suddenly find themselves with no more pressing concern than to stuff themselves, in preparation for another long winter's "nap." Some of them bed down in May or June, and by mid-July, when the weather is still warm and food still abundant, all the adult males are snugly underground. At most, their aboveground season spans four months.

The females, in the meantime, have their families to raise. On the average, the young are born about the beginning of May, seven or eight to a litter. They are naked, deaf, blind, and so helpless that, if rolled onto their backs, they cannot right themselves. Yet by the time they take their first timid peek above ground, four or five weeks later, they have hair, teeth, hearing, sight, and can recognize members of their family, probably by scent. Their first attempts to stand upright at the mouth of the burrow often end in pratfalls down the hole, but otherwise they're steady on their feet and can frisk up and down the tunnel at the slightest cause for alarm. By this time, they will already have increased their weight some thirteenfold (from six to eighty grams) and will more than treble it again by fall. This lively pace of development is characteristic of hibernating mammals, which have to pack a year's worth of eating, growth, and activity into the few short months that they're alert.

Most of the adult females (those that have

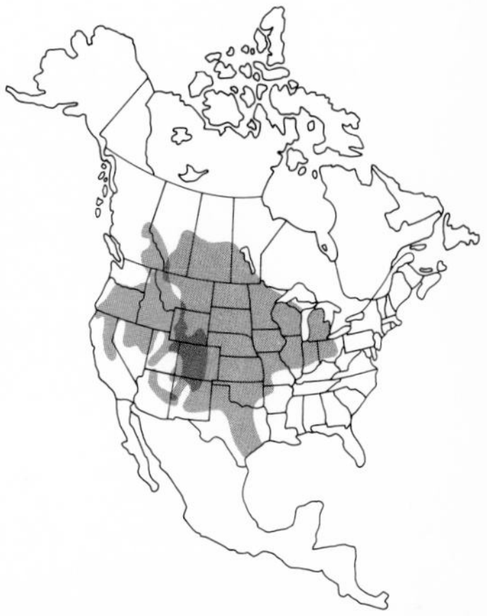

Thirteen-lined ground squirrel
Golden-mantled ground squirrel
Range overlap

Young Richardson's ground squirrels usually make their first timid ventures above ground in late May or early June. This species is most likely to be confused with Franklin's ground squirrel (not illustrated), which is larger and has a long, bushy tail, rather like that of the tree squirrels.

borne and weaned young) are fat enough to go underground by late July, but the youngsters may continue eating until October or November. The last stragglers above ground are always immature males, an ill-fated lot, since as few as three percent of those present in an area during the summer will show up there the following spring. What becomes of all the rest? Some may simply take off for parts unknown, either in July, when the young males tend to disperse, or during the breeding season. But many must surely die. It may be that, because they get last pick of the hibernation sites, they often have to settle for second-rate winter accommodation and hence freeze to death. Or perhaps their travels and extended period of activity make them especially easy marks for hawks, weasels, badgers, coyotes, foxes, snakes, tame cats and dogs, and all the other creatures that feed on ground squirrels. Of course, predators and other perils take their toll on the rest of the population as well, with the result that well under half the adults may survive from one year to the next. For Richardson's ground squirrels, four years is a ripe old age.

While this high death rate keeps the population in check, local overcrowding might still be a problem if the squirrels weren't naturally programmed to keep themselves evenly spaced out. When a female Richardson's ground squirrel and her nearest female neighbor are active at the same time, they keep their distance, staying about twenty meters apart if they're close relatives and further yet if they're not. Should two animals share the use of an area, they take pains to visit it at different times. But if a confrontation does occur, the outcome is never in doubt: whichever animal makes more use of the spot that is in dispute will chase the other away. Each adult has a certain territory, usually around its main burrow, which it uses more often than any of its neighbors and in which it is dominant over all comers. In the outlying reaches of its

Facing page: As straight and fat as a bowling pin, this Columbian ground squirrel adopts the look-out posture that is typical of all western Canadian ground squirrels.

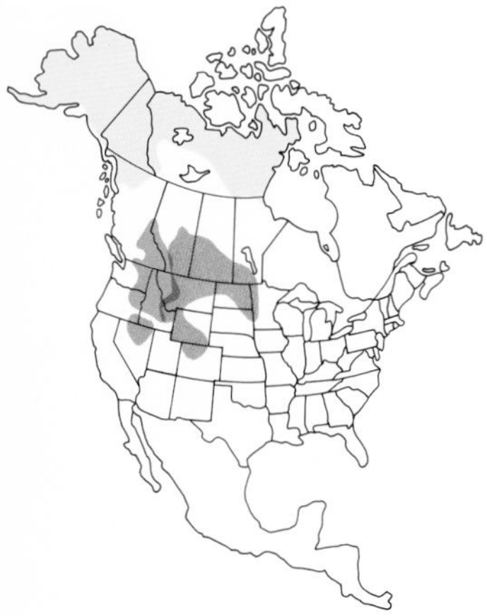


home range, where its activities overlap with those of its neighbors, it is less confident and can be routed by a squirrel that has a better claim to the area. Females have an additional psychological edge over males, who, though larger, are also more submissive by nature. Although males defend the vicinity of their burrows, elsewhere their overriding impulse is retreat, particularly when spurred on by the females' teeth. If you see a Richardson's ground squirrel with a scarred backside, it will probably be one of the down-trodden males.

Quite a different situation is found among arctic and Columbian ground squirrels, *Spermophilus parryii* and *Spermophilus columbianus,* in which males are dominant. These two highly social species (the former found on the tundra, the latter in the mountains and foothills) live in compact villages, or colonies. Though the females defend their burrows against other squirrels, most of the border skirmishes involve ambitious males. Serious fights among members of the community are rare, but sometimes marauding "toughs" from neighboring areas come in on raids. Among Columbian ground squirrels, these bullies, always large, dominant males, aim their attacks at the yearlings, who usually have enough speed and "street sense" to take refuge underground. But in the arctic species, youngsters of only a few weeks are mauled and sometimes killed. It is conceivable that this behavior benefits the species by encouraging the terror-stricken younger generation to disperse, thus preventing the colony from becoming cramped. At the same time, it is possible

For obvious reasons, the thirteen-lined ground squirrel is often known as the "striped gopher."

that it is an aberration, an example of predatory urges gone awry.

While there is no evidence that the raiding squirrels eat their victims, ground squirrels are, at times, carnivorous. This will hardly come as news to anyone who has watched Richardson's ground squirrels dining on a former associate that got in the way of a car, but it might have been a surprise for whoever gave the ground squirrels their scientific name, *Spermophilus,* or "seed loving." Of course, ground squirrels do eat seeds — up to a bushel each per year in the case of Richardson's. Yet they also like leaves, flowers, roots, insects, and carrion, among other things. The thirteen-lined ground squirrel, *Spermophilus tridecemlineatus,* for example, a species that is characteristically found on well-grazed native prairie, has a passion for grasshoppers. The Cascade golden-mantled ground squirrel, *Spermophilus saturatus,* which occurs only in western Washington and southwestern British Columbia, relies on both fruits and seeds. Golden-mantled ground squirrels, *Spermophilus lateralis,* those striped charmers that fatten on handouts in the mountain parks, naturally subsist on fungi, with leaves making up a substantial portion of their diets and seeds farther down the list. And Franklin's ground squirrels, *Spermophilus franklinii,* (which range from the north-central states, up through southern Manitoba, and into the aspen parklands of Saskatchewan and Alberta) are also catholic in their tastes. Their menu includes vegetation, insects, amphibians, and birds. They have even been known to prey upon young cottontails!

**This golden-mantled ground squirrel, sunbathing on a rock, might be mistaken for a chipmunk were it not for its larger size and the lack of stripes on the sides of its face. The Cascade golden-mantled ground squirrel is larger yet, and its markings, though similar, are relatively indistinct.**

# Red Squirrel

If you take a walk in a forest anywhere in western Canada, you may hear the chattering call of a tree squirrel. In the southwest corner of the British Columbia mainland, this sound announces the presence of Douglas's squirrel, *Tamiasciurus douglasii* (which also occurs in the Pacific states); everywhere else in western Canada, it is the voice of the red squirrel, *Tamiasciurus hudsonicus,* a species that is similar in appearance and habits. Like Douglas's squirrel, this little creature seems to delight in retiring to the safety of the treetops to deliver an impertinent scolding to all those animals, big or small, that dare to venture into its territory.

In reality though, most of the red squirrel's calls are directed at its own kind and serve much the same function as bird song — to advertise occupancy of a territory and to warn other members of the species not to trespass. These territories are all-important to red squirrels because, within its domain, each individual lays claim to its yearly food supply, stores it for the winter, and defends it from all other squirrels. Given a choice, a red squirrel prefers to stake its claim where it can obtain an abundant supply of its favorite food, the seeds of spruce trees. The squirrel spends much of each summer industriously harvesting its crop of seed-bearing cones. With boundless energy it works the treetops, nipping off the cones, sometimes bombarding the forest floor below with a shower of these missiles. If it seems to be in a hurry, perhaps it's because it has a deadline: in the fall, those spruce cones still remaining on the trees will open and shed their seeds to the wind. Before that happens, a typical squirrel will have gathered an incredible 14,000 cones and cached them in its underground tunnels, under logs, or in tree cavities, places where the dampness prevents the cones from opening.

One might expect that, since spruce-cone production varies greatly from year to year, red squirrel populations would fluctuate in response. This is true to some extent, but squirrel numbers are also regulated by the animals' territorial natures — generally, each squirrel has a plot of ¼ to 1½ hectares, an area that is usually large enough to contain a year's supply of cones even when cone production is low. If the spruce-cone crop fails completely, the squirrels still may not be caught short: in good seasons they hide away more than a year's supply of cones, accumulating an excess for use in leaner times.

Nor are red squirrels totally dependent on spruce cones. In some coniferous forests, many or all of the trees may be pines, and though the squirrels prefer spruce cones, the cones of some pines offer an important advantage — they are designed to unlock their seeds only after fires, not every autumn, so they are dependable, year after year, as a source of food. When cones of all descriptions are in short supply, red squirrels rely more heavily on other foods, the winter buds of conifers, for example. Some squirrels eke out an existence in stands of deciduous trees, relying on hazelnuts, rose hips, and the buds or catkins of alder and birch. If conifers are nearby, they will move to a better neighborhood when the opportunity arises.

In any habitat, red squirrels reveal themselves as true gourmands of fungi. While many mushrooms are eaten fresh, the red squirrel has a remarkable habit of placing others on logs or in the low branches of trees to dry. Except in wet climates, these delicacies are excellently preserved for winter consumption and may be left in the trees or relocated to dry storage areas, such as hollow stumps.

Caches of fungi, cones, or other food are usually located near the center of the squirrel's territory, close to a favorite hummock or stump where, with forepaws and teeth, the squirrel strips the tough outer scales off the cones to get at the seeds. At the squirrel's feet,

**Using its teeth and delicate "hands," this red squirrel dissects the seed cones from an alder twig.**

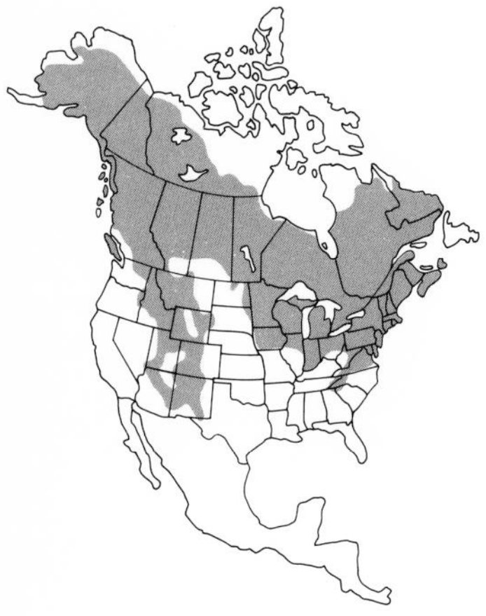

Red squirrel

the inedible portions of the cones may gradually accumulate in a huge refuse heap, or midden, which is added to by generation after generation of squirrels. A large midden may be seventy years old and measure over a meter in depth and five meters in diameter, an impressive monument to the industry of these tireless cone-shuckers. In northern Canada, and sometimes in the south as well, red squirrels convert their garbage heaps into winter insulation by constructing tunnels and nests inside them. As winter descends on the forests, the squirrels spend more and more of their time in their runways under the snow, jealously guarding the midden and food stockpiles. Unlike the ground squirrels, red squirrels don't hibernate, but air temperatures below minus twenty-five degrees Celsius curtail all of their above-ground activities.

With the appearance of the first patches of bare ground in the spring, whether it be in February, March, or April, red squirrels begin their courtship activities. About a month later, at a time when many changes in territorial ownership occur and when boundary disputes involving madcap chases are frequent, a female will declare a single day of general amnesty, allowing the neighboring males into her territory to quarrel over the opportunity to mate with her.

About forty days later, a litter of three to five young are born in a summer nest which is sometimes in a ball of twigs, moss, and grasses, high in a conifer, or less often inside a hollow tree. By the time the young are weaned, at about nine weeks of age, their mother may have set them up in separate housekeeping in another nest near the periphery of her territory, carrying them there in her mouth, one at a time, by their bellies. By fall, the young may claim a portion of the mother's territory, or they may disperse.

A red squirrel, especially a young one, loves to play by itself, engaging a leaf or a nut in mock combat or rolling and tumbling on the ground. In the trees, it can perform like a trapeze artist, swinging on a branch, falling and catching itself on a limb, or hanging by its hind feet. It seems to defy gravity, running upside down along a branch or leaping two or three meters between trees. Occasionally it slips and falls, saving itself from serious injury by parachuting with legs and tail outstretched. The skills that it learns at play serve the red squirrel well when attacked by one of its predators, the red-tailed hawk, coyote, lynx, or goshawk. Familiarity with its territory gives a red squirrel a chance to outjump and outwit the only predator that is likely to give it chase in the treetops — the marten.

# Northern Flying Squirrel

A rodent that flies? In point of fact, the northern flying squirrel, *Glaucomys sabrinus,* does not fly so much as glide. Its aerial excursions are always downward runs, either from one tree to another or from a tree to the ground. Instead of wings, the flying squirrel is equipped with "gliding membranes," capes of loose, softly furred skin that extend from wrist to ankle on both sides of the body. For climbing or running, these flaps fold back out of the way, leaving the animal as quick and nimble as that other highwire performer of the coniferous and mixed-wood forests, the red squirrel.

The flying squirrel usually makes its home where the trees are spaced apart, leaving room for its aerial exploits. To become airborne, it simply scampers up a tree, bobs its head several times to get a bearing on its destination, and then leaps into space with its front

Two special adaptations of the northern flying squirrel are large eyes, for night vision, and gliding membranes, which drape neatly against the animal's flanks when not in use.

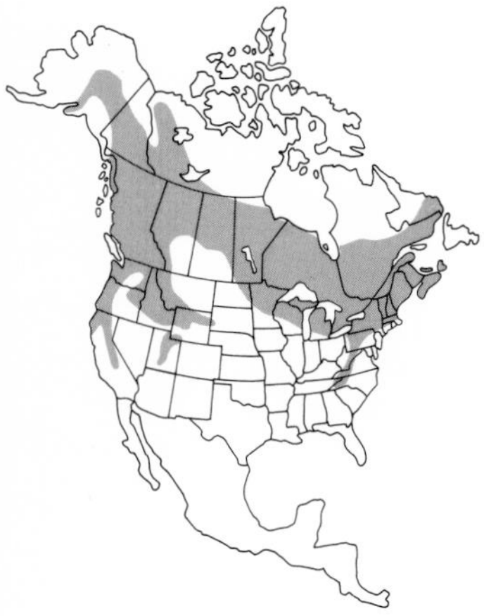

Northern flying squirrel

and back legs extended. This stretches the robes of extra skin that form the gliding membranes. Once aloft it can easily travel forty meters or more, swooping at will from side to side by manipulating its front legs. With the left leg low, for example, the animal veers left; by holding itself in turning position, it can even execute a tight downward spiral. These fancy maneuvers not only enable the flying squirrel to stay on target, something it does with surprising grace and precision, they also help it evade night-flying owls, which are its main enemies.

Like owls (and unlike all other North American squirrels), flying squirrels are nocturnal. Their active period is short, just a couple of hours after sunset and the last hour or so before sunrise; the rest of the time they apparently spend in their nests. These hideaways can take several forms. One common type is built in abandoned woodpecker holes or rotten trees. A good site is evidently passed down from generation to generation, for the remains of fourteen flying-squirrel nests, all insulated with shredded bark, were once found one atop the other inside a hollow tree. Flying squirrels do not hibernate, and in the wintertime, as many as nine squirrels have been found in this kind of nest, huddled together against the cold.

For the warmer months, the squirrels often build themselves airy, out-of-doors homes, perhaps beginning with the abandoned nest of a magpie, jay, or red squirrel. At their simplest, these shelters consist of bowls of twigs topped with balls of shredded bark, some fifteen to twenty centimeters across. These seem to be bachelor pads for adult males. Females with young put up more substantial lodgings — complete spheres of bark and twigs, about thirty centimeters across. Inside, these nests are insulated first with a layer of leaves and grass, then with a padding of finely shredded bark. Occasionally a mother squirrel shelters her family inside a big wad of hairlike lichens that she's picked off the trees. Whatever they're made of, such nests are usually located near the trunks of spruce trees, between one and ten meters up. If you happen to find one, look for others nearby, for an adult squirrel will often have several retreats. Besides, there may be another squirrel in the vicinity. It is common for the males (driven off by their mates just before the young are born) to take up residence near their families. At night, the adults feed and play together, but there is no sign that the father is ever permitted to help with the little ones.

The females bear three to six bright pink babies in late April or early May. The young are suckled for two or three months, a rather long infancy for a creature that, under natural conditions, has little chance of surviving beyond two years. The young aren't fully proficient at gliding until the end of their first summer.

Fortunately, their mothers are attentive guardians. The story is told, for example, of a northern flying squirrel whose nest and offspring were threatened by fire. Undaunted, the female picked up each of her infants in turn and carried them in her mouth to the safety of a nearby stump. Her coat was singed in the process, but her five offspring were unharmed. Had these youngsters been a little older, they would have helped their mother by grabbing hold around her neck for the trip to their new home. Thus encumbered, a female flying squirrel is able to run, climb, and even glide.

No one knows for certain what our flying squirrels eat, though their diet probably includes fungi, lichens, and other vegetation. They also have an appetite for meat which, regrettably, can be fatal: all too often, the squirrels are attracted to the bait in traps set for fur-bearing carnivores. These killings benefit no one since flying-squirrel fur cannot be marketed. So many of the animals are caught that trappers are often the best source of information on their probable whereabouts.

Many of us will never see northern flying squirrels in the wild. We are not at our best in the woods after the sun goes down. But if sometime you do have a chance to observe them carefully, be sure to take notes. What you've just read encompasses most of what is known about these appealing animals.

# Beaver

More than any other mammal except man, the beaver, *Castor canadensis,* is capable of altering its environment to suit its needs. Above all, it requires water, deep enough to provide escape from predators, to allow for safe, underwater entrances to its lodges and burrows, and to permit storage of its winter food beneath the ice. If this requirement is not met by a lake or river, the beaver dams a stream to create its own pond, cutting brush and felling trees for the purpose.

For an ax, this lumberjack carries two pairs of self-sharpening incisors that can fell a 20-centimeter-thick aspen in ten minutes. (Doubtless it took a little longer to bring down one tree on record, a cottonwood 170 centimeters in diameter.) Before getting down to business on one of its nocturnal tree-cutting forays, a beaver may stop for a snack, eating the bark around the base of the tree to be cut. Then, standing upright on its hind legs, head cocked sideways, it starts to gnaw through the trunk, making several bites before prying out each chip. The tree may fall in any direction, but only rarely is a beaver killed as it hurries away. With the tree down, the beaver begins cutting off limbs and dragging them with its teeth to the nearby pond or stream. As tree after tree is hauled piecemeal over the same route and obstacles are gnawed away or removed, a conspicuous tote-road is soon established. In low, marshy areas, such a trail

**A beaver heads for the underwater entrance to its lodge. Once inside, it will eat the bark off the stick it is carrying by manipulating it just as a person would an ear of corn.**

Beavers do most of their feeding in the lodge or in the water, safe from predators.

may connect with a canal, half a meter or more wide and up to 100 meters long. These waterways, excavated mainly with the front feet, are worth the effort that it takes to construct them because they allow segments of tree trunks to be floated out easily.

Tree-felling teeth are only one of the beaver's many special attributes. It has unusual, thick fur, so prized by trappers that, by the early part of this century, the beaver had reached the point of near extinction throughout much of its range. (Through the efforts of conservationists, it has since enjoyed a dramatic comeback.) By waterproofing this fur with castoreum, an oily secretion from two glands near the rectum, the beaver ensures that, when it submerges, its fur will carry along an envelope of air. This insulated diving suit, together with an exceptional lung capacity and flap valves on its nostrils and ears, outfit the beaver well for an underwater dive of up to fifteen minutes' duration. The large, webbed hind feet, which are used for propelling the beaver torpedo-like through the water, each carry two special split claws — a handy comb for grooming the fur.

For all-round usefulness, the paddle-shaped tail is an unsurpassed adaptation. The beaver employs it as a rudder when swimming, as a prop when tree-cutting, and as a counterweight when it walks semierect, carrying a load of mud or a young one in its arms. When alarmed, a beaver slaps its tail on the water's surface, warning other beavers to take safety in deep water. Brought forward under a mother's body, the tail may also serve as a heated nursing platform for the young. It is even useful when idle, by storing fat for the lean winter months. Some early writers, not noted for their honesty when recounting a story, reported even more remarkable uses for the beaver's tail. According to several accounts from the eighteenth century, beavers constructed their dams by hammering stakes into the streambed with their tails, interweaving a trellis of sticks, and then plastering the structure with mud, using their tails as trowels. In fact, a dam is constructed by piling brush, tree limbs, or other debris in the stream.

Mud, leaves, and rocks, mostly dredged up from the stream bottom, are deposited in armloads and serve to cement the jumble of sticks and brush. Long poles, dragged over the dam from the upstream side, are left with the upper end on the crest and the lower end in the mud below the dam. As more sticks and debris are added, the poles become embedded at their upper ends and act as props against the weight of the impounded water. Beavers are stimulated to build and repair dams, especially in the fall, by the sound of running water. Not possessing much reasoning ability, they will even attempt construction of a dam in a tub of standing water if they are played a recording of the sound made by trickling water. But in the wild, their innate responses to this and many other stimuli result in behavior that is almost as appropriate as if it were guided by reason. They don't, for example, build dams on lakes.

When circumstances warrant it, they don't even build their familiar lodges. If banks over a meter high are available, a beaver may construct a bank den, accessible only by underwater tunnels. The living chamber, about a meter in diameter, is above the waterline, often amongst the roots of a tree or bush. If a dam is built, as is usual on a stream, the rising water levels create problems inside the burrow. The beaver's solution is to renovate: it digs upward in the den, adding material from the roof to the floor. By gradually replacing the original roof with sticks and mud piled on from the outside, and continuing to dig and gnaw away from the inside, the bank burrow is eventually converted into a bank lodge. When beavers colonize an area without high banks, the lodge is constructed by heaping up mud, sticks, and other debris in shallow water, then tunneling up and into this mass from below the waterline. Armloads of mud are bulldozed onto the lodge, but the peak is left unplastered, allowing for ventilation. The completed dwelling, with a dry, spacious interior and two or more underwater entrances, is a fortress against the cold of winter and the attacks of predators such as wolves, coyotes, wolverines, and bears.

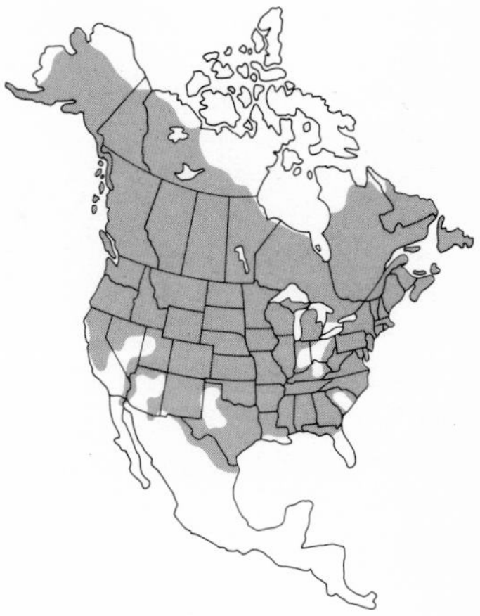

Beaver

All that is required to complete the beaver's living arrangements, in readiness for

High and dry inside the lodge,
a beaver nurses her kittens.

winter, is a nearby food stockpile. While grasses, leaves, water lilies, and other aquatic plants are important foods in summer, the twigs and bark of trees, especially aspen, are year-round staples and must be stored for winter consumption. Selecting an area of deep water, the beavers pile up branches of aspen, willow, alder, and other deciduous trees and shrubs, sometimes anchoring the butts of branches in the mud. Conveniently, most of this mass of brush soon becomes waterlogged and sinks below the level where ice will form. In winter, a hungry beaver has only to take a short swim under the ice to reach the food cache. It bites off a stick or a piece of bark and returns to the lodge or burrow to dine in comfort. Arctic beavers, even though their metabolism slows down in winter, often run out of provisions before spring and have to gnaw their way through the ice about a month before breakup to restock the larder.

Spring brings a sudden change to the colony. Over winter, a successful colony includes two adults, their yearlings, and their two-year-olds. But the two-year-olds bear the burden of insuring that the species spreads into new areas: in spring they disperse over land and water to establish new homes and new colonies. This family breakup is accomplished with a minimum of hard feelings. Placid by nature, beavers seldom fight, and when they do, it rarely comes to more than a hiss and a slap with a paw. Dispersing two-year-olds and members of neighboring colonies avoid conflicts by keeping out of occupied territories, each of which is marked by several scent mounds, little piles of mud and debris up to half a meter high, on which castoreum is deposited.

A few weeks after the departure of the two-year-olds, the colony acquires a new litter of kits, usually three or four little balls of fur that are soon scrambling around wide-eyed. Life is easy for the first summer — the kits give little help in the nightly chores of the older colony members and seldom even venture onto land, preferring to feed on the cuttings left in the water by others. They continue to grow throughout life and can achieve a record weight of forty kilograms (about half the weight of a man). But even these rare individuals are mere kittens compared with an extinct beaver from the Pleistocene period, a leviathan by the name of *Castoroides ohioensis,* which may have achieved a weight ten times this amount.

# Deer Mouse

No wild animal has received more human attention than the humble deer mouse, *Peromyscus maniculatus.* One reason for this interest is the expanse and diversity of the species' range. What other mammal is able to thrive on sagebrush prairies and at the treeline, inside a house and in deep coniferous woods? Although their staple foods in all habitats are seeds and arthropods, their menu varies from place to place, according to what's available. They can do just as nicely on conifer seeds and spiders, for example, as on beetles, weed seeds, and wheat. Their housing arrangements are also variable. As long as there's room to build a ball of grass and fluff about ten centimeters across, any nook or cranny suits a deer mouse fine, whether it's in an underground burrow, up in a bird's nest, or in the toe of an old rubber boot. To put it succinctly, deer mice are opportunists: versatility is the key to the species' success.

This is not to say that you can take an individual from one location and expect it to do equally well in another habitat. Each population of deer mice is adapted to the conditions in its local environment. Thus, mice that live in sandhills wear tawny coats that make them almost invisible, while mice that inhabit forests are exceptionally dark. (This camouflage is all the more effective since deer mice are nocturnal.) Forest-dwelling races are also equipped with extra-long, prehensile tails that help them keep their balance while climbing trees. Deer mice may also be adapted to their surroundings by their temperament. In the wide open spaces of the prairies, for example, where a tasty rodent is often exposed to the glare of predators, life puts a premium on being cautious and jittery. Mice from more sheltered environments, by comparison, are docile and relaxed.

Not only do deer mice respond to the long-lasting features of their habitat, they are also sensitive to such short-term variables as weather and food supply. For example, the length and timing of their breeding season varies from year to year; it might run from May to August or from January to July, depending on conditions. Even their mating behavior is subject to change. After a hard winter, during which most of the mice have died, the few surviving adults apparently indulge in a season of unbridled sexual promiscuity. But when more of the population makes it through the winter, the mice keep their numbers in check by setting up a social

**With its large eyes and soft, bicolored coat (tawny above, white below), the deer mouse really does resemble a deer.**

Deer mice generally live in small family groups; the animals shown here, huddled together for warmth, can be identified as juveniles by their dark gray fur.

Deer mouse

hierarchy in which only the "upper class" are likely to reproduce. Early in the spring, the heaviest and most belligerent of the males claim territories of about one hectare each. These are the breeding males; all the rest disperse, perhaps to establish homes somewhere else, but more likely to meet early deaths. The females, for their part, settle on home ranges of perhaps one-third hectare, spaced out so that there is little or no contact among them. Each breeding female does share at least part of her domain with one or more of the dominant males; these are her mates. Sometimes a single male and female associate as a monogamous pair.

Such mouse "marriages" frequently end with the death of one of the partners. Although the Methuselah among deer mice is reported to have passed away in its comfortable cage at the age of eight years, wild mice are generally snatched off in their prime. What with starvation, bad weather, and predators — badgers, raccoons, foxes, weasels, and owls, among others — only a fortunate few survive past eighteen months. Often a high percentage dies in infancy.

Interestingly, mice born late in the breeding season do much better than those born earlier on. In one typical study, only a quarter of the deer mice born in the spring and summer lived beyond five weeks, while the same fraction of the autumn litters was alive after five months. Biologists note that the rate of infant mortality is highest when the adult males are in peak breeding condition and, hence, at their most aggressive. In laboratory tests, juvenile deer mice that are forced into contact with combative males are sometimes killed outright. A few of them apparently lie down and die of shock; at the very least, their growth is stunted by the malign influence of the dominant males. Docile adults do not have this unwholesome effect on the young, so it stands to reason that in the fall, when the males come out of breeding condition and their tempers improve, the juveniles have a better chance to survive.

Not everyone is persuaded by this line of argument. Some researchers point out that breeding males often share their home ranges with nonbreeding subordinates, and no harm appears to be done. Occasionally, males even assist in the care of young litters. In one case when a mouse nest was disturbed, the parents cooperated in carting their chubby offspring to a new home. Then the male settled down to brood the youngsters and keep them from running away, while the mother, for some unknown reason, busied herself with eating and turning backward somersaults. Perhaps she was just excited at having a little time to herself: in one breeding season, a female deer mouse can bear up to four litters of three or four each; her life is taken over by maternal cares unless the male helps out.

In the winter, deer mice often congregate in companionable groups of mixed age and sex. There may be as few as three or more than a dozen huddled in a nest, most of them close kin. Although they have storerooms full of seeds, neatly sorted by kind, to last them through the worst of the weather, they go on foraging at night throughout much of the winter. At this season, the delicate tracery of their footprints where they've scampered over the snow bears testimony to their continuing activities.

# Bushy-Tailed Wood Rat

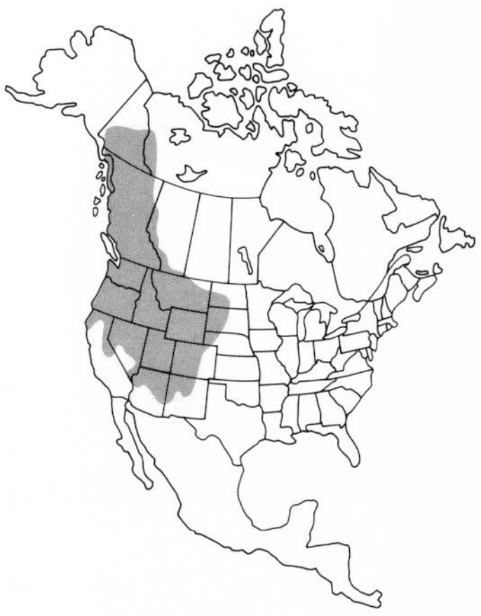

Bushy-tailed wood rat

Although you may not have heard of the bushy-tailed wood rat, *Neotoma cinerea,* the species' common name of "pack rat" is no doubt familiar. This epithet derives from the animal's well-known habit of making nocturnal raids into any human habitation within its range, pilfering small articles, and carrying them away to its rocky den. Although popularly supposed to prefer shiny objects, wood rats have also been known to make off with such varied items as dynamite and false teeth! Sometimes they leave bits of litter behind, as if in compensation for what they've removed, a custom that has earned them the nickname of "trade rat." These exchanges probably occur when rats that are out on collecting sprees drop what they're carrying in order to pick up something else that catches their attention.

The bulk of a pack rat's hoard consists not of man-made trinkets but of bones and sticks. Among the twenty-odd species of wood rats in North America, there are a number that use such materials to build their homes — piles of debris that sometimes reach a meter or more in height. Bushy-tails, by contrast, typically live where they can find shelter in rock crevices, caves, or abandoned buildings; very occasionally one will take up residence in a burrow, thicket, or hollow tree. Although it is not unheard of for bushy-tails to construct elaborate huts, such behavior is highly unusual, even among those sparse populations that have moved out of the mountains, along wooded river valleys, and onto the parklands and plains.

The mere fact that they don't have to build houses does not prevent the bushy-tails from collecting trash, and they sometimes amass heaps of sticks, bones, scats, feathers, and other junk for no discernible reason, unless it is simply to satisfy their acquisitive urge. More commonly, however, the debris is put to good use, in partially closing the opening of a rat's home crevice, for example, or providing a horizontal feeding platform for an animal that lives in a steep rock wall. In late summer and fall, the top of the pile may be used as a rack on which to cure leafy cuttings that are gathered for winter food. (Foliage from a wide variety of plants is the basis of the wood rat's diet throughout the year.)

In addition to winter fodder and building materials, wood rats also accumulate soft fibers for their nests. Each adult bushy-tail has one or more cup- or dome-shaped masses of grass and shredded bark, usually about twenty centimeters across, in which to rest. It is in these cozy structures, as well, that the females rear their families, which are born in spring or summer; one pregnancy per year is probably the rule at our latitude. Although a litter may include as many as six youngsters, the average number is three or four. This is fortunate, since the female has only four teats and her infants are compulsive about nursing. Within minutes of birth, each newborn creeps under its mother and clamps itself to a nipple (unless, of course, it arrives after all the positions are full). It will suckle more or less continuously during its first few weeks of life, maintaining a hold on its mother by means of its peculiar incisors, which flare out top and bottom to form a six-sided opening in which the teat can be grasped. This tenacious grip is probably more important in the nurslings' defense than in their nutrition, because, in the case of danger, the mother rat can flee with her offspring securely attached. Ever a nimble climber, she is capable, even when encumbered, of scaling cliffs or climbing trees to make her escape. Although litters of five are occasionally reared by wood rats in captivity, this accomplishment must be rare in the wilds, since the female can only protect four youngsters at a time, and predators abound: hawks, owls, carnivorous mammals, and rattlesnakes.

The young wood rats are weaned at about one month of age and strike out on their own a month or two after that. Once again, they find themselves in competition for a limited resource, in this case denning sites that offer sufficient shelter for overwintering. Good dens are used by generation after generation, or so one is forced to conclude from the quantity of satiny, black residue that builds up on the walls beneath the animals' favorite toilet perches — far too much for one rat to deposit in its three or four years of life.

Each individual usually has its own, private den. Females will generally tolerate other members of their sex in the neighborhood, especially if they are relatives, but the males are more solitary. Their goal in life is to control a territory from which all other males can be excluded. They are not so standoffish towards the females, of course, and each male mates with all females that live inside his realm (usually from one to three).

Although a "rat" in name, the bushy-tailed wood rat is not closely related to the Norway rat. With its plumy tail and silky fur, it looks rather like a squirrel.

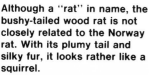

# Meadow Vole

When you see a hawk dive to snatch its prey from the ground or watch a fox or coyote hunting in a field, chances are good that the victim will be any one of several dozen species of mice and voles. Nearly all of our furred or feathered predators rely on these diminutive rodents for a portion of their food supply. Their importance in food chains, as converters of plant matter into little morsels of concentrated food, should be remembered before we condemn them for their habit of continually pilfering the farmers' crops.

One of the most common of these species is the meadow vole, or field mouse, *Microtus pennsylvanicus.* Like its many relatives, it does best in its own particular habitat — dense grasslands or low-lying, humid meadows. Here, in runways that form a criss-cross pattern throughout the miniature jungles of vegetation, it is sheltered from the eyes of predators. It also finds shelter from the elements in the dome-shaped grass nests that it weaves on the ground. Food is usually abundant in the form of insects, clover, alfalfa, horsetails, grasses, or other greenery. Other types of mice and voles, many of them very similar in appearance to meadow voles, share the same geographic areas by occupying slightly different ecological niches. Gapper's red-backed vole, *Clethrionomys gapperi,* for example, may live right next door to the meadow vole, in a stand of aspen or spruce. In the winter, these species remain active under the snow and may spill over into each other's preferred habitats, but in the spring, this cozy relationship is ended by mutual hostility.

We're used to thinking of mice and voles as being timid, but in their own little world, they can be surprisingly belligerent. Disputes often arise between meadow voles when one violates another's territory by coming within about three meters of its neighbor's nest. With the hair bristling on the nape of its neck, the angered proprietor may advance in a goose-stepping gait, hunched into a little ball of fury. The way it lunges at the intruder, stamping its delicate feet and gnashing its tiny teeth, would seem comical to a human observer, but in the Lilliputian world of field mice it must present a fearsome spectacle. If the intruder doesn't withdraw, the antagonists will do battle, either standing on their hind feet in order to box with their forepaws or tumbling on the ground and biting each other. Fights are most common during the reproductive period, when ninety percent of the adult males bear wounds from these encounters.

The females are less apt to fight, especially during the breeding period, when their energies are channelled into prodigious feats of reproduction. In captivity, one meadow vole once produced seventeen litters, of about five each, in a single year. Since successive generations of female offspring can breed at twenty-five days of age and bear their litters at forty-six days and then every twenty-one days thereafter, this one mouse could theoretically have produced over a million descendants by the end of the year. Fortunately, however, this potential is never realized in the real world, where only a portion of the females are pregnant at once and where their life expectancy is extremely short. Fewer than one meadow vole in fifty reaches the ripe old age of a hundred days, time enough for a female to bear only three litters.

Still, their capacity for rapid increase is apparent in their drastic population fluctuations, which recur every three or four years and are synchronized over large areas. One year, meadow voles can scarcely be found; a year or two later they may number 200 per hectare. A tenfold increase in a single year can be brought about by a lengthening of the breeding season and a lowering of the age of sexual maturity.

What causes the periodic decimation of the population? This is a question that many biologists have been trying to answer, with little success, for decades. About all that is

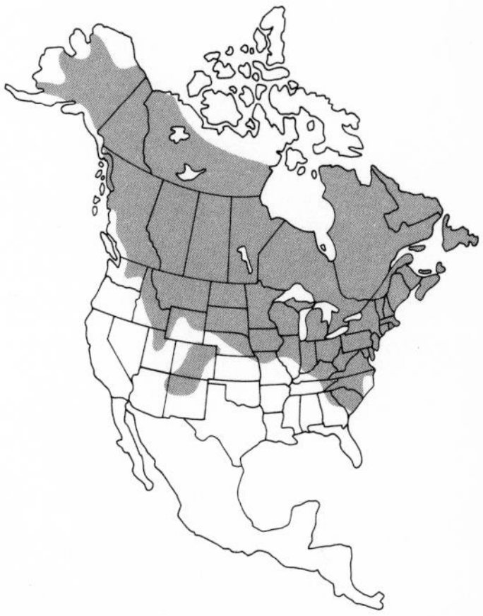

Meadow vole

known for sure is which factors are not involved. It isn't predation, nor a shortage of food, nor adverse weather, nor a contagious disease. Nor does it appear to be exhaustion or low resistance brought on by the stress of living in a high-density population.

The theory debated most often in the last few years is that the behavior of meadow voles changes during an increase in population, perhaps because different hereditary types are selected for. It has been suggested that when things start getting a little crowded, a docile, highly reproductive type disperses into less-suitable habitat, leaving behind those individuals that are more aggressive and hence more suited to high-density living, but that also have a lower rate of reproduction. The population of the stay-at-homes begins to drop, perhaps due in part to the failure of some pregnancies. But in may respects, this theory is still inadequate. It doesn't explain, for example, why the death rate increases during a population decline, or why some peak populations don't appear to be highly aggressive. It may be many years yet before the riddle is completely solved.

**Meadow voles are not often seen out in the open. Although they are active night and day, they usually remain hidden in the vegetation.**

# Muskrat

In spite of its name and long, scaly tail, the muskrat, *Ondatra zibethicus,* is not really a rat. From a taxonomist's point of view, it might more accurately have been dubbed "musk-mouse" or "muskvole," in recognition of its closest relatives. In many respects, the musk-rat can best be described as a one-kilogram meadow vole that has adapted to a semiaquatic life.

The muskrat is most at home in marsh-land, whether that takes the form of a thousand-hectare tract of cattails or the reedy margin of a drainage ditch. Ideally, water levels in a muskrat marsh are stable (prone neither to drought nor flood) and near the one-to-two-meter mark — deep enough so that there is free water at the bottom throughout winter, yet shallow enough so that aquatic plants receive the sunlight they need for growth. Here the muskrat harvests cattail and bulrush roots, arrowroot, horsetail, and underwater vegetation. As a rule, food from animal sources, such as fresh water mussels and clams, becomes important only when the supply of high-quality plants runs low.

As befits an animal that must dive after most of its food, the muskrat is completely equipped with underwater gear. Like the beaver, it has a "wetsuit" of glistening, waterproof fur that is insulated with a layer of trapped air. For flippers, it has partially webbed hind feet that are fringed with bristle-like hairs, and its laterally flattened tail serves as both rudder and oar. In addition, its fleshy, furred lips close behind its gnawing teeth so that it can forage underwater with its mouth "shut." It also has certain physiological adaptations that take the place of a snorkel or oxygen mask. For one thing, it is insensitive to levels of carbon dioxide in the bloodstream that would leave a human gasping. For another, it is able to withdraw oxygen from its body tissues when the store in its lungs runs out. As a result, a muskrat can easily go for two or three minutes between breaths and, by relaxing, can extend its limit to a quarter of an hour or more.

**Muskrat**

This ability is particularly useful after freeze-up, when the muskrat makes its living beneath the ice. Although shielded from winter's worst excesses of cold and storm, this sealed-off world suffers from a shortage of fresh air. The muskrat solves the problem by maintaining protected breathing places, locating them, on the average, forty meters apart, or about as far as it cares to sprint on a single breath. These refuges are of two types. "Feeders" are squat huts of rush and cattail,

**Though generally most active in the early morning and after sundown, muskrats can often be observed during daylight hours.**

less than half a meter high, with room inside for several runs into the water and a platform on which the muskrat can eat in comfort, high and dry. "Push-ups," or "breathers," are inconspicuous domes of frozen waterweed, mostly mealtime debris, that has been shoved together to cover a crevice or plunge hole in the ice. Sometimes they are built where an upwelling of marsh gas helps to keep the surface clear, but this isn't necessary, because a " 'rat" can easily gnaw up through thirty centimeters or more of solid ice. Since muskrats are creatures of habit, with customary feeding platforms, toilets, travel routes, and the like, it is not surprising that feeders and push-ups tend to appear in about the same places year after year. Often they lead like stepping stones to a bank burrow or lodge in which muskrats dwell.

The winter lodge, which may be home to more than a dozen animals, is the acme of muskrat architecture. Typically, it is built in the fall as the collective project of what is presumed to be a two-generation family group. The first step in the construction is the selection of a secure foundation, such as a mud bar or the roof of a collapsed lodge, on which to pile rushes, cattails, pondweeds, and muck. Ultimately, the heap will be more-or-less conical and about three-quarters of a meter tall. (The occasional oversized dwelling approaches two meters in height.) A burrow or two is dug into the mound from under the water, and one or more nest chambers are hollowed out inside. More material may be plastered on the exterior until the walls are eighty centimeters thick! When it's minus fifteen degrees Celsius outside, it may be plus fifteen within, particularly if the chamber is filled from floor to ceiling with slumbering muskrats.

Although these winter quarters are well insulated, their ventilation is poor, especially when the inner surface becomes glazed with ice. The nesting chamber may reek of hydrogen sulphide and other decay products, and by midwinter, oxygen levels often begin to fall as exhaled carbon dioxide builds up inside. Who could blame the muskrats for becoming restless as spring arrives, with its promise of fresh air?

The real physiological basis for this outbreak of wanderlust is the advent of the breeding season. At our latitude, female muskrats usually have two litters a year, for a total of about a dozen young, which they rear in bank burrows, lodges, or open nests. Although naked, blind, and apparently helpless, the newborn 'rats are exceptionally tough: they can withstand severe chills, recover from serious wounds, and survive for several days without food. These capabilities are a blessing since a female muskrat is not a model mom, and even when the male takes a hand in child care, as he often does, it does not make up for her deficiencies. Those infants that the mother does not lose, ignore, or trample, she may ultimately kill in her attempts to drive them off before the birth of another brood. Her philosophy of motherhood, if she had one, would be something to the effect that "life is cheap." And so it is, in a muskrat marsh. Muskrats may be eaten by mink, pike, and other predators; they are subject to devastating epidemics and recurrent climatic disasters; they are under steady pressure from human trappers. Yet such is the fecundity and resilience of the species that muskrat populations continue to flourish throughout North America.

# Porcupine

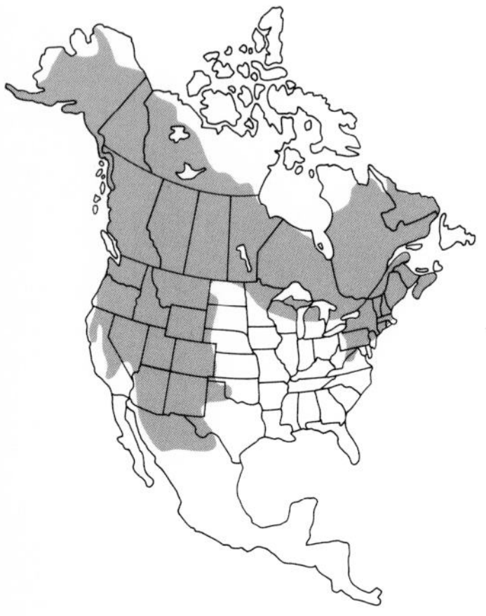

Porcupine

Of the 400 or so species of mammals that are now native to the United States and Canada, almost all are thought to have evolved on this continent or in Eurasia. Two exceptions are the opossum, which originated in South America and then dispersed north, and the porcupine, *Erethizon dorsatum,* which may have had a similar history. The frost-sensitive possum now sometimes ventures as far north as the lower Fraser Valley, but the "quill pig," more hardy by far, can make itself comfortable in nearly any forest or patch of brush south of the tundra. Ideally, porcupines settle in mixed woods, where they can banquet on such things as catkins, buds, and leaves in season and then switch to a diet of conifer bark for the rest of the year. When these preferred winter rations cannot be obtained, the animals make do with the bark of deciduous trees instead.

In one respect, the porcupine might be considered the giraffe of the North American woods, because it can browse in the treetops, far beyond the reach of animals like deer and hares that share some of its tastes in food. Slow and bumbling on the ground, the porcupine is a confident tree-climber, using its long, curved claws for traction and the bristles at the base of its muscular tail for extra support. It also has remarkable powers of balance and a nice sense of how far to press its luck as it treads on slender, swaying branches in the crown of a tree. It often seeks the heights, knowing that the most succulent greens can be found there; yet reports of a porcupine dying in a fall are very rare. Wherever you see bright, barkless scars well up in a tree, it is a sure sign that a porcupine has been around. You can also hope to see the animal itself, perhaps stretched out drowsily along a branch.

Another way of finding a porcupine is to locate a den, which can be recognized by the carpeting of brown, bean-shaped dung. Porcupines are easygoing about their housing requirements: they'll take shelter in hollow logs, windfalls, brush piles, culverts, old buildings, and abandoned cars, whatever they can find; some individuals regularly sleep out in the open, part way up a particular tree. In winter, they make caverns or tunnels in the snow. Generally speaking, they are not possessive about their accommodations. They add no amenities, not even so much as straw for a bed, and thus have little to lose if they seek other quarters or if another porcupine temporarily usurps the spot. Still, they evidently appreciate snug, secure lodgings when they can get them, because in a good denning area, one with numerous rock-walled crevices for example, they will sometimes congregate by the dozen. Yet even under such exceptional circumstances, they seldom get so chummy as to bunk two to a den.

Porcupines are a solitary lot. This raises the interesting question of how they find one another in early winter when it's time for them to mate. To date, no one can be certain of the answer, but they likely rely on scent. A porcupine fixes the world with a vacant, half-blind stare, but its nose is constantly a-twitch. (Its sensory powers are in keeping with its primarily nocturnal way of life.) We know that porcupines select their food by smell, scarcely seeming to see it when it's right before their eyes. We also know that the entrances to their dens and the bases of their favorite feeding trees, as well as the trails they follow in between, are often marked with the musky scent of porcupine urine, an odor so strong that it can sometimes be detected by people, despite our second-rate sense of smell.

To a sexually aroused male porcupine, the odor of the female's urine is the ultimate aphrodisiac. If he finds a place where a female has wet, he will sometimes snatch up a pawful of damp soil, hold it to his nose, and amble around on his hind legs as if savoring the bouquet. He may also ritually anoint the female with his own scent. For this to happen, both animals must stand erect and face to face.

Bucktoothed and bristling with about 30,000 quills, the porcupine is unique on this continent. All of its close relatives live in Central and South America.

After a little preliminary nose rubbing, the male lets loose a fountain of urine that wets his "*objet d'amour*" from nose to tail and sends droplets flying to a distance of two meters. The female responds somewhat frostily to this display of ardor, by fleeing, biting, striking out with her paws, or voicing complaint.

Although we generally think of porcupines as silent, they make a variety of sounds, ranging from mournful murmurs to catlike yowls. During the mating season, the male often gives vent to his feelings in a song that builds to a shrill, piercing whine as the animal becomes fully aroused. Whatever we might suspect, this vocalization is not due to the pain of embracing his prickly mate. Like other rodents, porcupines copulate by rear mount: the male's belly rests against the female's upturned tail. Since the undersides of their bodies are unarmed, no quills get in the way.

Sometime between May and July, after a pregnancy of seven months, the female gives birth to a single, precocious pup. It emerges with its eyes open and some of its teeth already in place, including the all-important, orange-colored incisors. Although its quills are soft at first, they quickly harden, and within five or six hours, the youngster is able to deploy its weapons in the classic defense display. With its backside turned to the source of the disturbance, the animal hides or lowers its head and then elevates its easily detachable quills, the better to offer any attacker a muzzle full of spears. If danger comes too close, the infant may lash out fiercely with its little tail. In a newborn, these tail-thrashings are ineffectual, but a swat from an adult can drive the quills to a depth of a centimeter or more.

Once in the flesh, the quills are held in place by microscopic barbs on their tips. These catch in the tissues of the victim so that any movement draws the quills into the body. If a porcupine gets pricked, in a fight, for example, it uses its teeth to yank out any of the quills that are within reach. An afflicted animal that lacks this skill runs the risk of infection or, if the quills interfere with foraging, of starvation. Preying on porcupines is thus a dangerous business, requiring much finesse, and, although many species of carnivores attempt it, the only ones to be consistently successful are fishers and mountain lions.

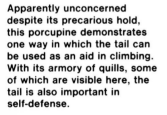

**Apparently unconcerned despite its precarious hold, this porcupine demonstrates one way in which the tail can be used as an aid in climbing. With its armory of quills, some of which are visible here, the tail is also important in self-defense.**

## Orders Odontoceti and Mysticeti

# Whales, Dolphins, and Porpoises

The cetaceans, or whales, dolphins, and porpoises, are the only mammals to have returned completely to the evolutionary womb of all terrestrial life — the oceans. To do so, they have had to make drastic modifications to the basic mammalian body plan. For propulsion they employ a powerful, horizontally flattened tail. For steering and balancing, they possess two flippers, which are highly modified forelimbs. The hind limbs are either entirely absent or, like the pelvis, reduced to a few vestigial bones, buried deep within the flesh. Hydrodynamic efficiency is achieved by a streamlined body shape: the neck is absent; nipples and sex organs are concealed within slits in the body; there are no external ears; and there is no hair, except for a few bristles on the snouts of fetuses and some adults. Partly in response to the problem of staying warm in frigid waters, cetaceans have developed thick layers of insulating blubber. Many of them also reduce heat loss by a high surface-to-volume ratio: freed from the limits to size that gravity imposes on terrestrial mammals, the whales have evolved enormous body weights, up to 145 tonnes for an exceptionally large blue whale.

Even in their most obvious link to their terrestrial ancestors, namely in their requirement for air, cetaceans show remarkable specialization. Instead of nostrils, they have valvular blowholes, usually at the tops of their heads. And many cetaceans, especially the large whales, can hold their breath during extremely long and deep dives. The record is held by the sperm whale, which frequently stays under for an hour or more and has been known to reach a depth of more than a kilometer. Such feats are made possible by several adaptations, including the presence of cartilaginous rings in the lungs, which keep small passages open and allow for rapid gas exchange. The animals' capacity to store oxygen is increased by having twice the usual concentration of red cells in the bloodstream and up to nine times the usual concentration of myoglobin in the muscles. The loose structure of the ribcage allows the lungs to collapse at great depths, thereby helping to prevent build-up of nitrogen, which can result in the "bends." And during a dive, the heartbeat slows and blood is redirected, bypassing nonessential organs.

Due to the efficiency of their many adaptations to an aquatic environment, cetaceans have been able to populate the world's oceans and fill a great variety of niches. Their success is reflected in their diversity. There are approximately eighty species (about a half of which occur in the coastal waters of western North America), broadly classified into two orders (or suborders, in the opinion of some scientists): the toothed whales, or Odontoceti, and the baleen whales, or Mysticeti.

The toothed whales include some of the great hunters of the oceans: the energetic dolphins and porpoises; the giant, squid-eating sperm whale; and the arch enemy of many marine mammals, the killer whale. They differ from the baleen whales in having teeth; in having single, instead of paired blowholes; in usually having well-developed echolocation ability; and in being generally more gregarious. They are also a larger and more varied assemblage of families, including four that are represented off our coasts: the beaked whales, or Ziphiidae; the sperm whales, or Physeteridae; the narwhal and beluga, or Monodontidae; and the killer whale, dolphins, and porpoises, or Delphinidae. Of the many species included in these families, the killer whale and the porpoises have been selected for detailed discussion in this chapter.

The baleen whales embody one of nature's ironies: these giants sustain themselves by feeding on zooplankton and small fish, some of the tiniest animals of the oceans. This is accomplished with the aid of hundreds of baleen plates, which form a huge, comblike structure that hangs from the upper jaw, in place of teeth. Each "tine" in this massive "comb" is a triangular, horny slat, fringed with bristles on its inside edge. When feeding, a baleen whale scoops up a continuous flow of water or gulps immense mouthfuls, filtering out the edible component on its mat of bristles. The baleen whales fall into three families, all represented in our waters: the gray whale (family Eschrichtiidae), having grooves on the outside of the throat and lacking a dorsal fin; the rorquals, or "finners" (family Balaenopteridae), having both throat grooves and a dorsal fin; and the right whales (family Balaenidae), distinguished by the absence of throat grooves and by the lack of a dorsal fin. The gray whale and the humpback whale (a rorqual) are two of the most familiar and easily identified species; both are discussed in the following chapter.

# Killer Whale

**Distribution:** Though found in all oceans, killer whales are most numerous in the colder waters of both hemispheres, right to the edge of the polar pack ice.

From the Arctic to the Antarctic, all around the globe, the oceans of the world are home to the killer whale, *Orcinus orca*. At weights of up to nine or ten tonnes, the adults are among the largest predators that have ever lived. Mighty yet lithe, they can jump straight up, clearing the water, or cover ten to fifteen meters in a single, arching leap. Even encumbered by the weight of a full-grown sea lion held crosswise in the jaws, they have been known to "porpoise" lightly through the air. Swift (with a maximum speed of around fifty kilometers per hour), ever-active, and intelligent, they are at the top of the food chains wherever they occur.

Killer whales can be identified in the wild by their piebald markings and tall dorsal fins. Nowhere can the animals be more easily observed than off the coasts of Washington and British Columbia. More than 200 of these so-called "blackfish" are known to frequent the enclosed waterways from Puget Sound to northeastern Vancouver Island, some as residents, others as visitors. It has been estimated that an additional hundred or so occur elsewhere along the British Columbia coast. Present year round, they are most abundant from late spring to fall and are particularly common in inshore waters during salmon runs. They evidently move about as necessary to satisfy their appetites, which call for 50 to 100 kilograms of fish, squid, and other food each day. Their travels are not without limit, however, for preliminary evidence suggests that each group of killer whales may be largely confined to a particular home range. "J pod," for example, the most frequently observed group of killer whales on the west coast, lives exclusively within the waters of Georgia Strait and Puget Sound, where it swims continuous clockwise circuits of its feeding grounds.

By studying J pod and its neighbors,

researchers have gained insight into the private lives of killer whales. Thanks to a patient analysis of several thousand whale "mug shots," in which individuals can be identified by subtle differences in markings and body contours, we now know that these "wolves of the sea" usually live in stable groups of from five to twenty animals (possibly extended families). Typically, about one-fifth of this number consists of mature bulls, which can be recognized by their overall length (up to ten meters) and the height of their dorsal fins (1½ meters or more). Another three-fifths or so consists of slightly smaller individuals — the females and subadults — while the remainder are young calves. At times, the cows and immature animals travel in tight formation, with the calves in the center and outriders at the rear and on both sides. The bulls may hold themselves aloof from this protective phalanx, sometimes by as much as six or seven kilometers, but they evidently keep in touch, since the entire pod maintains the same course.

The most likely medium for long-distance contact among killer whales is sound. They emit at least two different kinds of signals: clicks, which are thought to be used in echolocation, and screams, which probably function in communication. Just how attentive the whales are to one another's calls is suggested by the case of a female that was lassoed at sea, in preparation for towing her to an oceanarium. As she struggled to escape, the cow broadcast her misfortunes in strident tones, and within twenty minutes, a male had appeared at her side. Together, the two animals proceeded to slam and thrash the offending boat, with such effect that they were both killed for fear of a shipwreck. (Killer whales have, on rare occasions, sunk small vessels, usually for good cause, though never once have they attacked the escaping mariners. In fact,

**Elegant in its symmetry, the tall dorsal fin of the killer whale is one of the characteristics by which the species can be identified.**

This behavior, which has been given the apt name of "spyhopping," is common to many species of whales, including the killer whale, shown here. In addition to the markings that can be seen in this photograph, killer whales typically have patches of white extending from the belly part way up the flanks.

they are not known to have killed a human being under any circumstances.)

The fellow-feeling among killer whales is documented by a number of other reported incidents. Researchers working in Johnstone Strait, British Columbia recently had the exceptional experience of witnessing a killer whale birth. Throughout the labor and delivery, the mother (who was traveling in a group composed of two pods, her own and another) was closely accompanied by several females, who periodically paused to mill about and vocalize, as if in concern. Then, while the rest of the group waited at a distance, the cows moved into a shallow cove, where, amid continuing signs of excitement, the calf was born and lifted to the surface for its first breath. Shortly thereafter, with the group again united, all the whales participated in a wild performance that included spyhopping (see picture), splashing, vocalization, and raising the calf into the air. When calm had been restored, the animals resumed their travels, with the infant flanked by a bull and a cow — but not the one that had just given birth. Presumably, this gave the mother a chance to recover or deliver the afterbirth.

On another occasion, a killer whale was struck and apparently wounded by the propeller of a boat. Immediately, two other individuals positioned themselves on either side of the invalid, holding it up where it could blow. Incredibly, this assistance is thought to have been tendered for at least fifteen days.

The most impressive demonstrations of cooperation amongst killer whales occur on the hunt. Although many observations involve relatively small hunting parties (probably single pods), coordinated attacks by larger "superpods" (temporary aggregations of several groups?) are also on record. Once, during a salmon run, forty killer whales were watched as they apparently herded the fish inside an ever-diminishing circle, beginning with an area several kilometers in diameter, then patiently, over a period of hours, closing the perimeter. When the corral was only about a kilometer wide, the whales began to feed, taking turns as attackers and herders. This technique, with variations to suit the circumstances, is also used against other schooling prey such as herring and dolphins.

Another common hunting strategy among killer whales is to ram the prey from below, delivering a blow that sends it flying into the air. This method is used in the polar oceans to dislodge penguins from ice floes and walrus pups from their mothers' backs. In the Pacific, it is sometimes employed against sea lions and harbor seals. Killer whales have been known to follow this attack with a tail-lashing that rips hunks of flesh off their stunned victims. At times they will "play" their prey before killing it, catching and releasing and catching it again, like a king-sized cat with a giant mouse.

The killer whale's most remarkable feat of predation is the killing of "great whales," an accomplishment matched by no other mammals apart from people. Attacks — not all of them successful — have been reported on gray, minke, humpback, and blue whales, among other species. Although details vary from case to case, the assault may begin with bites to the quarry's flippers, flukes, or fins. As the pursuit continues, the hunters sometimes box their victim in on all sides, preventing its escape and interfering with its attempts to surface for breath. Periodically, a squad may lunge underwater to tear away blubber and flesh, but this violence is not always required. A minke whale that was attacked by killer whales off the British Columbia coast died without apparent bloodshed, and it was later determined that it had likely drowned. By the time the carcass was recovered for study, it had been stripped of its skin, an operation carried out with such delicacy that the underlying blubber was unmarked. As is usual when killer whales feed on other whales, the tongue and the flesh of the lower jaw had been devoured; the rest of the carcass had scarcely been touched.

Predation by killer whales on their own kind is almost unheard of. Indeed, the principal causes of death for this species in the wild are not well documented. Disease takes its toll, and rarely, a pod becomes stranded in shallow water. In addition, a comparatively small number is taken by human whalers.

# Porpoises

**Dall's porpoise**

There are certain cetaceans that seem too small to qualify as whales, so they are known instead as dolphins and porpoises. Human fascination with these clever and outgoing creatures has spanned the centuries. Even today, many people look to them for evidence of human-like language and thought, though majority scientific opinion has tentatively ranked them between the dog and chimpanzee in intelligence.

Of the half dozen species that occur off the coast of British Columbia, the two that are most frequently observed are Dall's porpoise, *Phocoenoides dalli,* and the harbor porpoise, *Phocoena phocoena.* The former, with its eye-catching black-and-white flanks, is amongst the most beautiful of sea mammals. Although widely distributed in the North Pacific, Dall's porpoises are most abundant to the north of Vancouver Island. In schools of two to six (rarely as few as one or as many as a hundred), they inhabit waters as far as 2,500 kilometers from shore. They are also commonly seen in certain broad, coastal seaways, such as Hecate Strait and Queen Charlotte Sound. A favorite of mariners, they love to race in the bow waves of boats — providing they can find vessels fast enough to make the sport interesting: thrill-seeking Dall's porpoises have been clocked at speeds of up to fifty-five kilometers per hour. While they do not ordinarily leap clear of the water, they often raise distinctive "rooster tails" of spume as they burst to the surface for air.

(About the only species, apart from the harbor porpoise, with which Dall's porpoise is likely to be confused is the Pacific white-sided dolphin, *Lagenorhynchus obliquidens,* which sometimes raises a similar pattern of spray as it slices the surface of the water with its tall, hook-shaped fin. To a knowledgeable observer, however, the white-sided dolphin is not difficult to identify, even from afar, since it is exceptionally gregarious: schools of from three to fifty animals are typical off British Columbia, and groupings of a thousand or more occasionally occur. White-sided dolphins are also notable for their skill as acrobats. Not only do they commonly leap into the air, they are the only cetaceans in the North Pacific that naturally include a complete somersault amongst their aerial stunts.)

Compared to most other dolphins and porpoises, harbor porpoises are retiring and sedate. Their subdued coloration is unobtrusive, and they generally keep a safe distance from humans and boats. They are most commonly glimpsed in sheltered, inshore waters, usually in groups of two to five. When cruising in calm seas, they surface to breathe with a series of smooth, unhurried rolls, raising neither a spout nor a splash. All that is likely to give their presence away is the soft "huf-uhhh" of their breath, a sound that has earned them the nickname of "puffing pig."

This name is doubly appropriate because a healthy harbor porpoise is very fat. An average individual measures about 1½ meters in length and weighs some fifty kilograms, making it the smallest of marine cetaceans. Because of this small mass and streamlined form, a porpoise faces special problems in keeping itself warm. Hence the need for extra insulation: the harbor porpoise is wrapped in a blanket of blubber that makes up forty percent or more of its body weight. As in other cetaceans, this layer is equipped with "heat exchangers" — pairs of blood vessels so arranged that venous blood, returning to the heart, is warmed by the outgoing arterial flow, thus reducing the heat that is lost to the environment.

This conservation measure is not sufficient in itself, so both Dall's and harbor porpoises have accelerated rates of metabolism. This accounts for their keyed-up dispositions and hearty appetites. A harbor porpoise may be on the go day and night, pausing frequently for brief periods of rest. In twenty-four hours, it will consume about ten percent of its body weight in food. A Dall's porpoise, though somewhat larger and more robust, can be counted on to eat as much or more, since it has a relatively thin coat of insulating fat.

Both Dall's and harbor porpoises feed on small fish and squid; because of the peculiarities of their dentition, neither species is capable of handling large or heavily armored prey. Dall's porpoises are particularly eccentric in this respect, since their teeth rarely protrude out of the gum tissue. Instead, they have horny bumps on their gums, one between each pair of teeth, which permit them to grip their food. In harbor porpoises, the teeth are visible but very small — about a quarter of a centimeter in diameter. Not surprisingly, therefore, prey is swallowed whole. It then passes into the first compartment of the porpoise's three-chambered stomach, where it is broken apart before being digested. Sometimes a porpoise unwisely attempts to gulp down more than it can swallow, as in the case of one animal that choked to death on a small shark.

Herring appears to be a staple food for both species, though it is by no means their only resource. The muscular Dall's porpoise, for example, plunges down a hundred meters and more in search of hake and other deep-dwelling fish. And off California, several hundred harbor porpoises were once observed as they feasted on a huge school of Pacific sardines. While a number of the porpoises circled the fish to keep them from dispersing, others formed lines of five or six individuals which then lunged through the school in crescent formation. As each group completed its pass, it dived and swam under the sardines to keep them at the surface, while another squad of attackers rushed at the fish.

There are no other detailed reports on the feeding behavior of these porpoises in the wild. However, laboratory studies suggest that both Dall's and harbor porpoises (like other toothed whales) can hunt and navigate by sonar, using a finely tuned biological system that is far superior to its manmade counterparts. By emitting rapid-fire bursts of echolocation "clicks" and then interpreting the reflected sounds, a harbor porpoise can home in on a school of fish from a distance of one kilometer. Combinations of clicks are also thought to be used in communication — to produce the bleat of fright, for example, and the variety of grinding and squeaking sounds that accompany courtship. Unfortunately, harbor porpoises are so "soft spoken" that no study of their signalling has been possible in the open ocean.

Harbor porpoise
---International boundary

Dall's porpoises can be identified by the distinctive white patches on their flanks and by their darting speed. As youngsters in particular, they like to race ahead of boats and in the pressure waves of migrating gray whales.

# Gray Whale

Major summer concentrations of the gray whale
— Migration route
--- International boundary

In winter and early spring, from almost any headland on the west coast of North America, an animal migration on an immense scale may be witnessed. Nearly the entire world population of gray whales, *Eschrichtius robustus,* passes in review as they voyage from Arctic seas to the coast of Mexico and back again — a 16,000- to 22,000-kilometer journey that is unsurpassed by any other mammalian migrator. Most of the whales hug the coastline closely, except when passing indentations or major waterways such as the Strait of Juan de Fuca or Queen Charlotte Sound. Alone or in pods of up to sixteen individuals, they travel day and night, seldom stopping to play or feed. Surfacing and blowing four or five times in a minute, then raising flukes for a dive of about four minutes' duration, their steady rhythm of swimming and breathing measures the distance.

Between late February and early May the processions bear northward towards summer feeding grounds in the Bering, Chukchi, and Beaufort seas, off the western and northern coasts of Alaska. In these cold, shallow waters, the main food of the gray whale abounds — small, bottom-dwelling crustaceans, a centimeter or two in length, known as amphipods. In contrast to other baleen whales, the gray whale is a bottom feeder, ideally suited to exploiting this rich food source. After diving to the bottom, a feeding gray whale rolls onto its flank, protruding the lower lip on its deeper side; then with the aid of the powerful tongue, it vacuums in amphipods, water, and considerable quantities of mud and sand. A thrust of the tongue expels the prodigious mouthful, trapping the edible portion on the coarse, inner fringes of the baleen plates. Before moving to a new feeding spot, the whale surfaces to blow, often with clouds of mud billowing from its enormous head and lips.

Most gray whales feed mainly in these northern seas, then, during a six-month-long southern sojourn, enter a nearly complete fast and lose twenty to thirty percent of their body weight. But there are exceptions to this general pattern. While very little food is found during the migrations, gray whales have the ability to profit from a variety of food supplies should the opportunities arise. For example, pods of three or four migrating grays have been observed circling around and beneath schools of small fish, thereby herding them into tight groups, then taking turns at rising through the center of the school with their enormous jaws agape. Other exceptions to the pattern of feeding in northern seas and then fasting on the southern migration are found among those gray whales that pass the summer far south of the main population; a group of forty, for example, summers off the west coast of Vancouver Island, feeding mainly on an extremely abundant type of bottom-dwelling tube worm.

Regardless of where gray whales spend the summer, they head south in December and January, many of them bent on returning to their ancestral calving grounds — warm, shallow lagoons off the coast of Baja California, the most famous of which is Scammon's Lagoon, named after a nineteenth-century whaler. Mating usually occurs en route but may also take place in the calving lagoons or other southern waters. Considering that adult gray whales measure eleven to fourteen meters in length and weigh fourteen to forty tonnes, mating is a remarkably frolicsome affair, accompanied by breaching, spyhopping (raising the head perpendicularly from the water, presumably to have a look around), fluke standing (raising the tail perpendicularly from the water), and a good deal of rolling and thrashing about. Usually, copulation involves an hour-long ménage à trois, the third participant being a male whose function is possibly to stabilize the mating pair and keep them pushed together. Single calves are born, usually in February of the following year, after a thirteen-month gestation period. They mea-

"Fluking up," a gray whale commences a dive that will last for several minutes. White lines that show on the upper surface of the right fluke are scars from a killer whale attack.

sure about five meters in length at birth, and on a milk diet that is more than half fat, they nearly double in length by the time they are weaned in August. They are sexually mature between five and eleven years of age but continue growing until they are about thirty.

A major impediment to a gray whale living out its potential lifespan of fifty years or more is the killer whale, the gray whale's only natural predator. One out of every five grays bears the scars of an attack by a pod of killer whales — damaged tongues or tooth marks on flippers and flukes. In an attack, killer whales apparently attempt to drown the gray, perhaps by holding it underwater; the gray's best defense is to seek refuge in kelp beds or very shallow water.

The gray whale has also to contend with whaling ships. Because of the species' narrow, predictable migration path, its slow swimming speed, and its habit of congregating in calving lagoons, the gray whale is easily exploited and by 1900 had reached the point of near-extinction. Since 1947, however, when seventeen nations signed the International Convention for the Regulation of Whaling, the gray whale has enjoyed almost complete protection. (Some are taken for scientific purposes, and under the terms of the treaty, the government of the Soviet Union kills about 160 grays each year, in northern waters, for use by its aboriginal peoples.) Under this protection the gray whale has made a dramatic comeback and today numbers about 12,000, a figure that is probably near the level that the species maintained before commercial whaling began.

**With its snout up out of the water, this gray whale calf displays the mottled coloration that is characteristic of the species. One of its paired blow holes is also visible here.**

# Humpback Whale

Many an ancient sailor, becalmed in a silent sea, must have thought he was hearing some frightful ghost when eerie calls began to penetrate faintly through the wooden hull of his vessel. In many instances, the "ghost" would have been the humpback whale, *Megaptera novaeangliae,* the most vocal of the baleen whales and one that is found in all the oceans of the world. This species formerly numbered approximately 100,000, but today there remain only about 7,000 protected by international agreement since 1966. Of these survivors, several hundred frequent the west coast of North America. Other members of the rorqual family, such as blue, sei, fin, and minke whales occur in this area, but the humpback can be distinguished from them by its pear-shaped blow, about two meters high, and by the exceptionally long, narrow flippers, usually white on their undersides, which are well displayed in relatively frequent, spectacular breaches. The humpback whales in this area, like many other baleen whales, are seasonal residents; they summer off the coast, from California to Alaska, and winter in the warm waters surrounding the Hawaiian Islands and bordering Baja California.

It is mainly on the wintering grounds that the haunting songs for which the humpback is famous can be heard. Each performance is a six- to thirty-minute-long solo of some of the loudest and strangest sounds produced in nature. Whines, shrieks, moans, trills, snores, and utterances that are best compared to the creaking of a giant hinge or to the scream of a passing train whistle are strung together in repeating patterns and sound rather like the music of an avant-garde composer. All renditions by whales in the same area are of the same song, except that lengthy passages are often omitted. Uncanny in their ability to memorize and recite such long and complicated sound sequences, humpbacks also display an apparent talent for composition — the song changes subtly week by week, until, by the end of the season, the old song can scarcely be recognized in the new one.

Why do humpbacks sing? Respectable biologists are not likely to suggest that humpbacks, like people, perform for aesthetic reasons, but other explanations seem inadequate to account for the length, complexity, and variability of the songs. Some of the explanations that have been proposed suggest that singing may allow individuals to recognize one another and maintain group contact in an environment where visibility is restricted or, conversely, that the singing is a form of territorial advertisement, which serves to maintain spacing amongst individuals, much the same way as it does in birds.

One possible function of humpback songs is sexual attraction, since mating, like singing, is largely confined to the wintering grounds. Most mating occurs between October and March, accompanied by courtship behavior that includes breaching, lobtailing (slapping the surface of the water with the flukes), rubbing the bodies together in long glides, and exchanging affectionate strokes or resounding slaps with the flippers. Single calves (very rarely twins) are born in the same waters a year later. They are weaned at about eleven months of age, shortly before their mothers commence their next pregnancies. A female humpback is a devoted and gentle parent: if danger threatens, she may guard her young one under a fin or guide it with her snout; if a calf is injured, sometimes even if it dies, she will support it so that its blowholes clear the surface of the water.

The mother also leads the calf during its first migration (and possibly subsequent ones) to and from the summer feeding grounds. From Baja California, humpbacks of all ages and sexes head north along the coast in March or April, passing Vancouver Island in May or June. Not all of them complete the two-month journey to the Gulf of Alaska, the Aleutian Islands, and the Bering Sea; instead many

**Major concentrations of the humpback whale**

Facing page: Thirty-five tonnes of breaching humpback hurl skyward. What function such behavior serves is unknown — it may be for advertising ownership of a territory, for dislodging parasites, or it may be simply play.

summer in less northerly areas. The route taken by the large population of humpbacks that migrates from Hawaii remains a mystery, but their destination is presumably the same northern seas.

Their 5½-month stay in these cold waters, after a season of fasting in warmer, less-productive seas, is a prolonged feast by comparison. In the most common feeding technique, known as "lunge feeding," the whale surges forward, open mouthed, and plunges into the midst of its prey — a school of fish, such as anchovies, herring, sardines, cod, or salmon or, more likely, a concentration of krill (small shrimplike crustaceans). Pleats that extend from the whale's throat to its navel first expand, as the cavernous mouth engulfs both water and prey, then contract, as the living soup is strained through the baleen plates.

There are several variations on the basic technique of lunge feeding. For example, the whale may use the flukes to splash water forward, over its head, to momentarily startle and confuse a school of krill upon which it is advancing. Or it may extend the long, wing-like flippers to herd prey into the path of its gaping maw. But the most sophisticated trick that the humpback employs is likely the "bubble net." To set its net, a humpback dives beneath a school of fish or krill, then releases bursts of bubbles as it rises in a spiral around the school. The curtain of rising bubbles scares the prey into the center of the ring just as the whale surfaces, in the middle of its trap, to collect a concentrated mouthful of food. Is this an innate behavior pattern that the whale uses without understanding, or is it an indication of an advanced capacity for reasoning? No one knows the answer to this and many other intriguing questions about humpbacks.

Below: The humpback calf usually swims just above and to one side of the cow, gaining protection and, as is shown here, an occasional nudge to the surface for a breath of air. The throat grooves are a characteristic of many species of whales, including the rorquals, but the long, winglike flippers are unique to humpbacks.

## Order Carnivora

# Carnivores

According to the common definition, a carnivore is an animal that eats flesh. How can it be, then, that there are flesh-eating mammals, such as shrews, that are obviously not carnivores? And conversely, how can there be undoubted carnivores, such as grizzly bears, that eat little flesh? The answer to this riddle lies in the fact that "carnivore" has not one, but two meanings, which overlap. On the one hand, it can refer to any creature that consumes animals for nourishment, whether that creature is a pitcher plant or a wolf. On the other hand, it can be used to indicate only mammals that the taxonomists, guided by their knowledge of evolutionary relationships, have included in the order Carnivora. It is in the second, more restricted sense that the word will be used in this chapter.

The carnivores originated about 60,000,000 years ago as a group of small, forest-dwelling mammals called the miacids. One of the distinguishing characteristics of these primitive predators was their teeth. Like their insectivorous ancestors, the miacids possessed incisors, canines, and cheek teeth, for nipping, piercing, and crushing, respectively. Unlike their forebears, they also had shearing teeth, or carnassials, that permitted them to slice through muscle fiber and tough sinews. These specialized cutting blades developed through modification of the last upper premolar and the first lower molar. The presence of carnassial teeth in this position persists in carnivores to this day.

These details of dentition, technical though they may seem, may have meant the difference between success and oblivion for the miacids. There was another group of ancient, predatory mammals, called the creodonts, that also possessed carnassial teeth, but theirs evolved out of molars farther back on the jaw. This may have left the creodonts short of the crushing and grinding teeth that are necessary for masticating plants, thereby making these animals less flexible in their diet than their miacid competitors. It may be partly as a result of this shortcoming that the creodonts are long extinct, while the miacid line survives.

There are about 250 species of terrestrial carnivores in the world today, the exact number being subject to dispute. Taxonomists are agreed, however, that there are seven families. Two of these, which include such animals as the mongoose and hyena, are restricted to the Old World; the other five are represented in western Canada, as well as in neighboring parts of the continent. The Canidae, or dog family, for example, includes our coyotes, wolves, and foxes. Some 50,000,000 years ago, their ancestors left the forest to hunt on the open plains, an occupation that requires both endurance and speed. Over time, therefore, the evolving canids acquired a runner's easy stride and long-legged build; they also moved permanently up onto their toes to become more fleet. But with limbs that are highly specialized for running, the canids lack dexterity, making it difficult or impossible for them to hold large prey with their paws and deliver a precisely oriented killing bite, as a cat might do. On the other hand, several canids working in concert, each one attacking as it has the chance, may be able to bring down animals larger than themselves. (One concomitant of this tendency towards cooperation is the sociable nature of many canids, including "man's best friend.") In addition to meat, many members of the dog family also eat vegetable foods, so it is not surprising that the canids retain a full complement of teeth, with molars for pulping up plants, as well as sharp incisors, powerful canines, and well-developed carnassials. To make room for this toothy arsenal, they typically have long jaws, which give them their characteristic "doggy" snouts.

The Ursidae, or bears, are a comparatively late offshoot from the canid line. With their flat-footed gait and heavy limbs, they are specialized not for speed but for strength, an attribute that enables them to claw into fallen logs and overturn rocks in their search for food. Except for polar bears, the ursids tend toward vegetarianism, a fact that is reflected in their teeth: a bear's molars are enlarged and its carnassials poorly formed. A similar situation is found in the Procyonidae, another family of omnivorous carnivores, which is represented in western Canada only by the raccoon.

The Mustelidae, or weasel family, have moved toward the opposite extreme. Rather than emphasize crushing teeth at the expense of carnassials, they typically come equipped with efficient shearing blades and a reduced number of molars, a pattern of dentition that is in accord with their "bloodthirsty" tastes. The mustelids are generally low-slung, forest-dwelling types, not unlike their miacid ancestors. Most of them possess an offensive scent that they can exude in self-defense, a capability that is most notable in the skunks. Other members of the family include the marten, fisher, mink, wolverine, badger, weasels, and otters.

The fifth and final family to be considered includes the Felidae, or cats, a group of proficient hunters that eat almost nothing but meat. Accordingly, they have deadly canines and large carnassials but have lost several of their grinding teeth. This helps account for their powerful, shortened jaws and rounded snouts. Unlike the canids, the felids probably arose in broken or forested country, where prey can often be taken through stealth. Thus, they have retractable claws that can be drawn in while traveling, for a quiet footfall, and extended when needed to hold prey. Although generally not adapted for long-distance pursuits, cats have the ability to manipulate their prey and attack it with precision. This means that, typically, they can hunt and kill alone. Correspondingly, most felids — from cougars down to house cats—are relatively unsociable.

Compared to the members of most other orders of mammals, the carnivores have never been abundant. This is an inevitable result of their role as predators: the hunter can never expect to outnumber its prey. But few though they are, they have received a generous measure of attention from human trappers, hunters, and pest-control officers. Certain species, notably the swift fox and black-footed ferret, are believed to have been exterminated from the western Canadian part of their range. Most of our other carnivores have been reduced in both numbers and range, and one species, the sea otter, is now classified as imperiled.

# Coyote

Over the millenia, the coyote, *Canis latrans,* acquired an exalted place in North American Indian myth. A number of tribes cast this intelligent wild dog in the hero's role of trickster-creator, and the great Aztec nation, from whose language our word "coyote" derives, linked the species with several of their deities. Amongst these, for example, was Coyolxauhqui, the goddess of the moon, whose association with the coyote no doubt stemmed from the animal's habit of howling high-pitched serenades into the nighttime skies. With the arrival of European settlers, however, the coyote quickly lost its nimbus of divinity.

"Standing-over" is a common component of both play and aggressive behavior among coyote pups and other young canids. More often than not, the "top dog" in these encounters is, in fact, the dominant animal.

Many people came to view it as a "bad animal" because it sometimes competed with humans for livestock and game. As late as the 1960s, one respected Canadian biologist could still revile the entire species as "knavish, furtive and cowardly, mean and treacherous." Given this attitude, it is no surprise that the coyote was persistently set upon with poison, traps, and guns.

The results of this onslaught have not been encouraging to the coyote's foes. Although certain local populations have been wiped out and total numbers have probably been reduced, the species continues to thrive. Ironically, people have unintentionally assisted the coyote by opening forested areas and by eradicating large predators such as cougars and wolves. Both these interventions have allowed the species to expand its range dramatically since 1850. Once confined to the area between Alberta and southern Mexico, the "little wolf" now occupies open country and broken forest throughout seventy degrees of latitude.

Wherever they live, whether in the tropics or on the arctic coast, coyotes fill much the same ecological niche. Although willing to eat almost anything that moves and some things that don't (including watermelons!), coyotes are basically scavengers and predators of small mammals. Unlike wolves, which specialize in killing large ungulates and hence hunt in packs, coyotes usually concentrate their attentions on such prey as lagomorphs, rodents, and domestic sheep. Thus they can hunt alone or in small parties. In remote areas, where the animals are not molested by man, coyotes may form permanent packs, but apparently these rarely function as hunting groups. Instead, their primary function appears to be the defense of ungulate carcasses, taken as carrion, which are a preferred source of winter food, where available.

Coyotes vary their hunting techniques to suit their prey. Even a pup of six or seven weeks, whose only previous achievement has been killing insects, may be able to catch itself a mouse. Alerted by a rustle and scurry in the grass, the infant hunter will instinctively assume a perfect "pointing" pose, a position that, as an adult, it will be able to hold with utter concentration for ten minutes or more. Then suddenly, it will leap into the air and jackknife down onto its unsuspecting quarry. In the winter, coyotes listen for rodents under the snow and then dive in, sometimes plunging in so deep that only their tails poke out.

To capture rabbits and hares, coyotes at times rely on speed: flat out, they can run at about fifty kilometers per hour. But even this is not enough to guarantee success, so the hunters occasionally try to improve their chances through teamwork. For example, one member of a pair will sometimes mill around inside a likely looking thicket while its partner patrols the edge, on the lookout for fleeing hares. A similar tactic may be used against ground squirrels, with one animal digging into the tunnel while the other waits beside an exit.

A more complex strategy may be necessary for killing a young fawn or elk calf, not because of any resistance that the prey puts up, but because of its maternal bodyguard. Occasionally, coyotes will down a feeble doe or buck, but a healthy adult ungulate can usually ward off attack with its death-dealing front hooves. This defense is all the more formidable if, as often happens, a number of outraged guardians present a common front. Alert to the danger, a pair of cooperating coyotes will sometimes split up before approaching their quarry so that one of them, the "decoy," is out in the open, while the other is hidden from sight. The decoy's job is to provoke the babysitting adult or adults into giving chase. If this ruse succeeds — and it doesn't always work — the second coyote appears out of nowhere to kill the undefended youngster, usually with bites to the head and neck.

Coyote hunting-partnerships may be either temporary or permanent. Whenever two adult coyotes are consistently seen hunting as a team, they are very likely mates. This is true regardless of the season, since some pairs stay together the year around. Each female comes into heat sometime during the winter, usually in February or March, and may be trailed for days or weeks by a mob of eager males. The

Coyote

suitors jockey amongst themselves for position, each trying to be first in line, but it is the female that ultimately decides which one of the front runners she'll accept. If she has bred before, she often gives precedence to her mate of the previous year, with the result that coyote mate-relationships tend to be long-lasting. Judging from the tail-wagging delight with which a female greets her partner after they've been apart or from the loyalty with which he regurgitates food for her while she's nursing their pups, the bond between a pair of coyotes is cemented with affection.

Both the male and female share in rearing the young, which are born in April or May; an average family numbers five or six. The pup's first home may be any one of several inconspicuous dens that the mother has prepared in advance, sometimes with the father's help: if the burrow gets so deep that the female can't push the dirt out with a single shove, the male may crawl in behind her and take up second position in the bucket brigade. The natal den is usually on a shrubby bank, but it may also be inside a brush pile, in an old building, under tree roots, or in abandoned badger digs. If the family is disturbed by people in the first month or so, one of the parents (usually the mother) will cart the pups to a new home. Alternatively, this task may be undertaken by a "nursemaid," an unmated coyote of either sex that has attached itself to the family and assists with feeding and defending the pups.

Among the more interesting aspects of the pups' upbringing are the communal howling sessions in which they participate. Such a chorus, which can be raised by night or day, begins with a long, sonorous wail from one of the adults, followed by the pups' piping renditions of the same phrase. These "singalongs" may serve to strengthen the family bond, for a group howl among adults appears to bring the animals closer together and reinforce their dominance relationships. Or perhaps the pups need to practice the skill of identifying individuals by voice, in preparation for the day when they will make contact with other coyotes over long distances: as adults, coyotes often rendezvous after locating one another through an exchange of howls. In the fall, after the pups are on their own, it is thought that their parents occasionally issue the call for a reunion, presumably so they can bring food to their pups. In those places where coyotes live in packs, howling also serves to announce the group's presence on a territory, a mode of advertisement that is practiced by wolves as well.

**Alert, clever, and adaptable, the coyote won the admiration of the North American Indians, some of whom regarded the species as "God's dog."**

# Wolf

Long an object of hatred and fear, the wolf, *Canis lupus,* has been shot, poisoned, and trapped to the point of extinction throughout much of its original range in Europe and North America. But its public image is changing. In Canada and elsewhere, the remaining wolf populations are sufficiently isolated that they seldom prey on livestock, and new information about their lives and habits is replacing the centuries-old misconception that the wolf is a fiendish killer of innocent people. (There are only two proven cases of wild wolves attacking people in North America, and neither resulted in any bodily harm). Much of the new sympathy for wolves stems from a realization that in many ways they are very similar to us. Like many primitive human hunters, for example, they combine a capacity for extended physical exertion with close-knit, cooperative societies, in order to prey on mammals much larger than themselves.

Wolf packs usually consist of ten or fewer individuals, most of them descendants of one original pair. A stable social order is maintained by a ranking system, or "dominance order," that is reinforced by the day-to-day interactions of pack members. In many ways they behave like dogs, which were domesticated from wolves over 10,000 years ago. When a subordinate wolf approaches a dominant, it adopts the original "hang-dog" expression, crouching with its ears laid back flat and its tail between its legs. In reply, the dominant assumes a regal manner, exactly opposite to that of the subordinate, walking stiff-legged, with ears pointed forward and tail held high. In a more intense encounter, the high-ranking wolf will attempt to keep the subordinate in

**An old deer carcass yields a few scraps for a pack of hungry wolves. Typically, they will make a new kill within a couple of days of abandoning the previous one, but if necessary, they can fast for two weeks or more.**

**Yawning, a wolf displays its weaponry. The two large teeth near the back of the jaw are carnassials, well suited for shearing sinew and crushing bone.**

line with a variety of ritualized threats such as staring fixedly, crouching as if about to attack, or growling with teeth bared. If sorely pressed, a subordinate may defuse the hostility of a dominant by acting like a puppy, rolling on its back and presenting its unprotected abdomen in helpless submission. The top-ranking, or "alpha" wolf is usually a male, often one of the original pair. He enjoys certain responsibilities and privileges of leadership: when he rests, the other wolves rest; when he attacks the prey, the others join in; when a kill is made, he may consume the choicest parts.

The privileges of rank are also exercised during the breeding season, in late winter; at that time, the alpha male and female are usually successful in their attempts to prevent others from mating. In May, when a litter of four to seven pups is born (most often in an underground den) the process of integrating the new members into the social order must begin. By the time they leave the den, at seven to ten weeks of age, permanent dominant-subordinate relationships may have been settled by playful or not-so-playful fighting with other pups. They will also have formed lifelong emotional attachments not only with siblings and parents, but with the other adults, all of whom participate in feeding the pups. By nipping at the muzzle of an adult, the pups stimulate it to regurgitate food. At the same time, the young ones are practicing some of the postures and behaviors that will later be used to express submissiveness.

**Wolf**

Food for the pups and for the rest of the pack consists primarily of whatever large grazing animals are most abundant in the area — elk and mule deer in the Rocky Mountain parks; caribou and muskoxen in the arctic; moose throughout much of the boreal forest. Where they are easily procured, bison, white-tailed deer, beaver, arctic hares, mice, even whitefish are featured on the wolf's menu. Since large kills are made infrequently, the wolf, of necessity, is something of a glutton in its feeding habits. At a carcass, a wolf tears off and bolts large chunks of meat and fat, occasionally downing as much as eight kilograms, or twenty percent of its own weight, at a single feeding. Usually a patch of scattered hair and large bones is all that remains after twenty-four hours, but if easily caught prey is abundant, the pack may abandon a partially consumed carcass to scavengers such as ravens, foxes, and wolverines.

The prowess that wolves display in feasting is matched by their stamina on the hunt. The pack generally travels fifteen to sixteen kilometers per day, locating potential prey by scent, by tracking, or by accidental encounters. When the prey is discovered, a life-or-death test of its fitness follows. Deer and caribou usually run or may ultimately seek refuge in water. Elk and moose are more likely to stand their ground, particularly if accompanied by a calf, and charge their attackers with flailing hooves. The wolves nearly always abandon the attack in a few minutes if the quarry stands and fights or demonstrates that it is a strong runner. Otherwise they attempt to sink their sharp fangs into the rump or flank of the victim and hang on, thereby bringing the animal down or at least damaging its running muscles. With smaller prey such as deer and caribou, or with a larger animal that has been weakened, the attack is also directed towards the vulnerable abdomen, throat, or nose.

Most of the time the prey escapes, perhaps to face the wolves another day. Those that succumb are often very young, very old, injured, or diseased. This selective predation thus has its beneficial effects: preventing sick animals from spreading disease; removing inferior animals from the breeding population; and in some cases, preventing populations from exploding, damaging their habitat, and subsequently starving en masse.

What it is that controls wolf populations remains poorly understood, but availability of vulnerable prey appears to be one important factor. When food is scarce, many of the pups starve, and the social system starts showing signs of stress. The wolves are harassed more by neighboring packs, which defend contiguous territories of 50 to 300 square kilometers or more.

A pack must also deal with the exceptional nonterritorial "lone wolves." These are de-

feated alphas or ostracized subordinates that travel widely, keeping a low profile at the periphery of pack territories, while they wait for the opportunity to find mates and start new packs. If a pack discovers a lone wolf or an alien pack in its territory, there is frequently an all-out fight in which one or more wolves are quite deliberately killed.

Territories are generally maintained by more peaceful means. By making regular patrols and urinating on scent posts, wolves keep their entire area, and especially the perimeter, marked with fresh "no trespassing" signs. They also voice their warnings to distant packs in choruses of eerie howls. A typical performance begins with a high-ranking wolf doing a solo, sometimes on a steady pitch, sometimes breaking to a higher note and dropping back. Others join in with yips, howls, and whines, for a minute or more, in what sounds like the melancholy lament of some forgotten spirit. To the wolves, however, it is just the opposite — a friendly tail-wagging ceremony that draws the pack into a close group.

These songs of the wild are most often heard on a still summer night and for a good reason. By that time the pups will have left the home den, but since they are not fully able to follow the adults until the fall, they remain at "rendezvous sites" while some of the pack is off hunting by night. In this situation, the pack uses howling not primarily for territorial advertisement, but to relocate each other after the chase or to draw together at the rendezvous site.

# Arctic Fox

Arctic fox

In the winter of 1962-63, a traveler arrived in North America after journeying from Russia on foot. The incident would probably have received wider publicity had the immigrant been a person, instead of an arctic fox, *Alopex lagopus;* nevertheless the fox's Russian neck tag informed a few people of the event. For such a small animal (arctic foxes weigh between 2½ and 9 kilograms) this is a remarkable feat, but long-distance movements across vast expanses of pack ice are not at all uncommon for the arctic fox. It is well equipped both for traveling and surviving in the realms of snow and ice that are found in Greenland and the polar regions of Eurasia and North America. Compared to other foxes, its limbs are shorter, its ears and snout more rounded, and its fur thicker, all adaptations to reduce heat loss. And in the winter its coat turns white, hiding the fox from the eyes of both predator and prey. (A "blue" winter color phase, in which the coat is dark bluish-grey to brownish, is common in Greenland and parts of Alaska but rare in western Canada, occurring in fewer than one percent of the population.)

Another of this animal's adaptations to life at the top of the world, where severe winters kill a large portion of the breeding population each year, is unusually large litters — perhaps the largest of any wild mammal. Ten or eleven whelps is average, and as many as twenty-five have been born in a single litter. Such a large family presents the mother with some unusual problems: one biologist was amazed to observe a vixen standing while her thirteen pups nursed with such vigor that, for a moment, she was lifted entirely off the ground by the muzzles of her young. Parenthood is demanding for the dog fox as well: from the time of the litter's birth, in late May or early June, until they are weaned five or six weeks later, he is solely responsible for supplying the vixen with food. She starts to share hunting duties with her mate around mid-July, about the same time as the pups first emerge from their underground dens to sit blinking in the sun or to romp about the den site.

Occasionally the den is simply a hollow under a large boulder, but most often it consists of a system of burrows dug in sandy

Dressed for winter, an arctic fox is oblivious to all but the coldest of temperatures. During severe winter storms, it may tunnel into a snowbank or search out one of the summer breeding dens.

soil, preferably on a south-facing slope. When the denning season begins, in spring, the foxes cannot dig in the frozen ground; instead they take up residence in already existing dens. A new burrow probably originates as a temporary shelter with only one entrance, excavated in late summer, when the ground is thawed down to the permafrost layer. With many generations of reuse and remodeling, it expands into a labyrinth of tunnels with as many as a hundred entrances spread out over hundreds of square meters. An old den is a very conspicuous feature of the tundra environment: as a result of droppings being deposited nearby, a large patch of lush vegetation springs up; and if the site is currently active, the tenants further advertise their presence with fresh piles of yellow sand, the result of their interminable digging. The average den sees more than 300 years of service before the last of the burrows finally collapses, gets dug out by a grizzly in search of ground squirrels, or is enlarged by denning wolves.

Most of the food brought to the vixen and her young at the den consists of lemmings, by far the most important item in the arctic fox's diet. It takes about a hundred of these morsels of prey, each weighing about as much as two meadow voles, to satisfy the daily requirements of a vixen and an average litter of weanlings. Birds and their eggs, when in season, are relished as a supplement to this fare. A single arctic fox has been known to plunder hundreds of goose nests in a few days, eating a small number of eggs and caching the rest by burying them. Other summer foods are mice and voles, fish caught in shallow water, caribou and sea mammals as carrion, insects, berries, grasses, and herbs. The garbage around human settlements and camps is a year-round source of food for this born scavenger.

Winter is generally a time of hunger for arctic foxes. Most of them leave their dens by September or October to wander widely in search of whatever meager fare is available. On the tundra, lemmings, which are dug out of the snow, are the mainstay in a diet that also includes arctic hare, ptarmigan, and caribou, the latter obtained by attacking weakened fawns or by scavenging at wolf kills. In coastal areas, arctic foxes regularly venture onto the bleak expanses of sea ice, often hundreds of kilometers from the shore. Here, they may subsist on seaweed and the carrion from sea birds and marine mammals, or follow polar bears to clean up their kills of ringed or bearded seals. The newborn seal pups are also frequently preyed upon by arctic foxes, which employ their extremely sensitive noses to locate the snow-and-ice birth lairs of the seals.

One of the unfortunate facts of life with which most arctic foxes must contend is a drastically fluctuating food supply. While North American lemmings do not periodically commit mass suicide by plunging into the sea, as popular mythology suggests, their numbers do fluctuate dramatically, usually by a factor of a hundred or more, every three or four years. The number of arctic foxes trapped for their furs reflects this cycle, the peaks in fur production regularly following a year or two after the peaks in lemming numbers. When the lemming population crashes, the number of breeding vixens may drop by nearly two-thirds, but many of them still breed, and with disastrous consequences. Faced with a critical shortage of food, a Cain and Abel war sets in amongst the sibling pups, resulting in the weaker being killed and sometimes cannibalized by their littermates. The stronger pups' victories are shortlived — the adults in the population prematurely abandon virtually every den, leaving the pups to starve.

Most of the adults face similarly harsh prospects. As winter sets in, they commence long-distance migrations in search of food. Their exodus follows natural routes such as coastlines or frozen rivers, often to areas far south of their normal breeding range. One of the white foxes has even been found within 150 kilometers of the Manitoba-United States border. A few begin moving back to their denning sites on the tundra in mid-February or March, but most of them never return. When lemmings are scarce, fatigue, cold, starvation, and trapping, as well as ever-present diseases like rabies and mange, all take their toll.

**Its winter coat mostly shed, an arctic fox sits on the mound of freshly dug earth near its burrow.**

# Red Fox

Red fox

A biologist studying the red fox, *Vulpes vulpes,* once observed a female, or vixen, pounce at a deer mouse but miss catching it. Unruffled, she withdrew to a nearby rise and lay down, apparently napping. When a rustle gave away her quarry's hiding place, the vixen pounced again — this time successfully. This incident gives some inkling of the intelligence and resourcefulness for which the red fox is famous. Anyone who studies this wily animal cannot help but admire these and many other appealing characteristics, including its graceful bearing and elegant appearance. Typically the animals are reddish-brown with white and black markings, the black being confined mainly to the backs of the ears, to the long, delicate legs, and to some of the guard hairs on the back and tail. But red foxes are not always red. A common variant in Alaska, the Yukon, and northern British Columbia, making up as much as one-quarter of the population in some areas, is the "silver fox," which is clad in black, except for a white-tipped tail and a sheen of white-tipped guard hairs. Another variant is the "cross fox," so called because it wears a dark stripe down the back and across the shoulders. It is also a cross in the genetic sense, for one way in which this type arises is through interbreeding between red and silver foxes.

The differences amongst these and other color phases of the red fox go more than skin deep. Fur ranchers have noticed, for example, that if you walk into a pen containing several individuals of different colors, the silvers will approach to investigate but the reds remain distant, their curiosity overcome by nervousness and fear. In the wild, the red fox is so shy and anxious that one study of a particularly dense population revealed that over half had ulcers, many of them acute.

In the lives of red foxes, a major source of stress is the neighbors. In areas where fox populations are not decimated by trapping or hunting, adult pairs usually maintain non-overlapping territories, typically a few square kilometers in extent. They stake their land claims in the usual canid fashion by urinating on prominent scent posts, and they confront or harass any trespassers. Territories are defended most avidly in the fall, at the onset of the critical winter period. In February, intraspecific relationships may be briefly strained again when the beginning of the breeding season provokes a retesting of the social hierarchy. On occasion there may be a serious fight, a whirlwind of fur, legs, and teeth as the animals battle in grim silence. Serious injuries can result, as evidenced by tufts of fur and drops of blood strewn on the battleground. Not surprisingly, ritualized forms of fighting have evolved, allowing for safer settlement of disputes. In one form of combat, the opponents balance on their hind legs, with their forepaws on each other's shoulders, while screaming and pushing at one another until one of them backs off or slinks away in defeat. A decisive battle can even be staged without physical contact: the two antagonists may simply sit or lie on the ground, almost muzzle-to-muzzle, for an extended screaming match.

However, quarrels are infrequent, especially in the spring and summer when food is more plentiful and raising of families is a major preoccupation. The pups are usually born in March or April, an average of five to a litter. At first they remain concealed in a spacious underground burrow that typically features a large living chamber, at least two entrances and, if the den is one of those that has been used for many years, a labyrinth of tunnels. It's easy to tell if a den is occupied — the entrances exude the characteristic odor of foxes and mounds of fresh, sandy soil testify to recent digging.

When the pups are about a month old, they begin to emerge for brief periods to play on or near these mounds of earth, struggling with one another and squabbling over their playthings — old bits of bone and feathers, leaves, or whatever objects are handy. Ever an

Screaming and pushing, two red foxes engage in a ritualized contest.

While a parent is off hunting, a pair of red fox pups enjoy the early morning sun outside the den.

anxious parent, the vixen usually oversees these antics. Should danger approach, her sharp warning bark sends the pups scrambling to the safety of their underground retreat. Parental nervousness may also explain why most litters are moved to new homes at least once, especially if the den has been disturbed. This habit of relocating the pups helps account for some of the instances where two different litters are found living together in the same den. "Communal denning," the living arrangement for about ten percent of the population, has some important benefits: in one case studied, both parents of a litter were killed, and a little while later the pups mysteriously turned up about half a kilometer distant, apparently living with the relatives — at least they were sharing a den with another family.

The pups are weaned at about two months of age, at the same time as they begin accompanying the adults on hunting trips, which usually occur at night. As their ties with the den site weaken, the center of activity for the family becomes one of a series of "rallying stations" within their home range. The male's role in raising the pups is poorly understood, but in some cases at least, he associates with the family throughout the denning and rallying-station period. The family breaks up in the fall: the parents go their separate ways until the next breeding season, and most of the youngsters disperse, usually no farther than fifteen kilometers but sometimes up to hundreds of kilometers.

By the time it is on its own, a young red fox is nearly fully grown and is well practiced in

Red foxes are seldom very active by day. They are mainly nocturnal hunters, with peaks in activity at dawn and dusk.

the hunting skills that it will need for survival. When hunting mice or voles, staples in its diet, a red fox relies on its superb sense of hearing. Searching silently, it hears something, then pauses motionless, ears pointed forward and its tail stiff and erect with excitement. A faint rustle is the prey's undoing. With a high, jackknifing leap, the fox pounces, attempting to bring its stiffened forelegs down sharply on the victim. If the prey evades the initial attack, the fox may frantically sniff and snap at the grass, pouncing repeatedly and flattening the concealing vegetation until it secures its prize. When stalking squirrels and birds such as pheasant or grouse, the fox crawls forward, catlike, keeping its belly to the ground and advancing on the target until a successful rush is possible. Rabbits, caught by surprise in their forms or after a short chase, are the main food for red foxes in southern regions. Farther north, where snowshoe hares take the place of rabbits, red fox numbers closely follow the snowshoe hare cycle. In the extreme north of its range, red fox numbers are tied to the three- or four-year lemming cycle.

Skilled as hunters, red foxes are also excellent scavengers. Nose to the ground, they pace about the forest floor, sniffing out insects, birds' eggs, dead animals, and other edibles. They keep track of their investigations by means of a highly efficient "bookkeeping system" — they urinate on inedible food remains, depositing a telltale scent that lasts for days and reads "no need to waste time investigating this again." The food sources disclosed by this searching may include the caches of other foxes. When they acquire a food bonanza, red foxes bury the surplus, bit by bit if necessary, in a series of shallow pits. Passing by the same way again, on a leaner day, the proprietor of these little hoards can unerringly remember the locations of the buried morsels, only a few of which may have been nosed out by other foxes.

# Black Bear

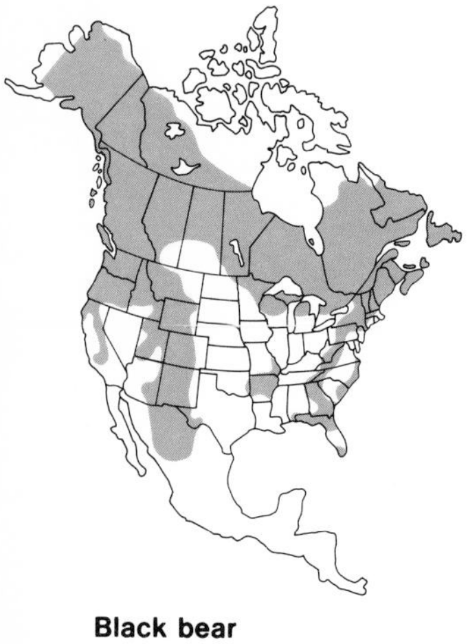

Black bear

According to one fanciful theory, the way to tell the difference between a black bear, *Ursus americanus,* and its close relative, the grizzly, is to sneak up softly behind the bear in question, give it a swift kick in the rump, then scramble up the nearest tree. If the bear climbs up after you, it's a black bear. There would, however, be more significance to this little drama if the roles were assigned differently: the black bear is, indeed, a tree-climber par excellence, but this ability very likely evolved to provide it with the means of escape from its larger relative, the grizzly. The black bear's shorter, more tightly curved front claws allow it to hitch itself up a tree trunk with astonishing speed, while grizzlies, except for the cubs, are almost completely confined to the ground. Accordingly, the black bear, at home throughout the forests of North America, seldom ventures far from the safety of trees.

Other than grizzlies, black bears have few natural enemies. Coyotes and wolves occasionally kill a young or hibernating bear, but the older male black bears have a much greater effect in limiting the population. An adult male may share its home range of thirty to fifty square kilometers with several adult females, but young bears and adults of the same sex are seldom tolerated. Aggressiveness amongst black bears is partly ritualized: for example, they select certain prominent trees in their territory to habitually claw and rub against. This results in heavily scarred "bear trees" that issue a visual and olfactory challenge to other bears. It is a challenge that is not lightly ignored, for small bears run the risk of being cannibalized by larger ones. Cubs are especially vulnerable, though they enjoy the constant care and protection of their mother. At the first approach of danger, a hoarse snort from her sends them up the nearest tree, and she either follows them herself or stands guard near the trunk. If, as is often the case, she is ready to breed by June or July of the cubs' second year of life, the hostility of her suitor breaks up the family. Female offspring are frequently allowed to establish themselves, temporarily at least, on their mother's home range, but young males are forced to disperse. Once they have lost the protection of their mothers, mortality rates soar amongst immature bears of both sexes.

Ultimately, though, the limits to black bear populations are set by the availability of food. In their feeding habits, they are best characterized as omnivorous opportunists — to put it bluntly, they will eat practically anything. Social insects, like ants and wasps, which can be lapped up in large numbers from colonial nests, are their main source of animal food, but they don't turn their noses up at fish, duck eggs, carrion, or even the occasional elk calf, caught bedded down. Their strength and resourcefulness leads to some highly unusual feeding methods. For example, some bears learn to locate trees containing the nesting holes of flickers or yellowbellied sapsuckers, probably by listening for the cries of the nestlings; then they climb to the nest and tear open the cavity with teeth and claws, to get at the hapless birds. Another unusual habit involves stripping the bark from the base of a pine, spruce, or fir tree, then using the incisor teeth to rasp away at the semisweet pulp on the exposed wood.

Vegetable food makes up most of their diet. In spring and early summer, black bears browse heavily on green vegetation. Horsetails are a favorite, and to get them a bear will wade belly-deep in a swamp, foraging like a moose in undignified abandon. In late summer and fall, the bears gorge themselves on hazelnuts, blueberries, cranberries, rose hips, huckleberries, and other wild fruits. Considering what pigs

**Black bears are usually true to their name, but blue-black, cinnamon, and even white individuals frequently occur. They are distinguishable from grizzlies by their smaller size, shorter front claws, and lack of a shoulder hump.**

they make of themselves during the berry season, when they may gain as much as fourteen kilograms in a week, it is appropriate that the adults are called boars and sows.

Their season of feasting prepares the bears for an extended fast. They enter hibernation about the time that food becomes scarce in autumn, and from then until spring, they consume no food or water (and pass no wastes), subsisting entirely on their bodily reserves. Compared to typical hibernators such as ground squirrels, whose body temperatures drop to within a degree or two of the freezing point, black bears do not reach a very deep torpor, thirty-two degrees Celsius being an approximate lower limit for their temperatures. But unlike ground squirrels, black bears do not regularly rouse themselves; instead, these Rip Van Winkles of the animal world remain continuously drowsy for as long as seven months, only rarely shifting positions or poking about out-of-doors during a warm spell. In central Alberta, for example, this dormant period extends from October or early November to early April. The den may be a natural cavity, such as a rock cave or a hollow under a stump or log.

Some of the females bed down for the winter with company — their cubs under a year of age. By late January or early February, other females may be sharing their dens with newborn cubs. There are usually two or three of the new arrivals, sometimes one, and very rarely as many as five. Born naked, blind, and helpless, weighing a mere 200 or 300 grams, their rapid development is a further drain on the body of their hibernating mother. By the time they leave the den, the cubs are already capable of following their mother on her travels, climbing trees for safety, and engaging in rambunctious contests of pawing, biting, and head-butting. When a little older, their play starts to include rearing up on the hind legs for brief cuffing matches with the front paws. Playing silently, they communicate their moods by a variety of facial expressions and ear postures. If the sparring starts to become objectionable to one of the combatants, the aggrieved party will lay its ears flat or give a low moan, signaling that the game must either end or escalate into a serious conflict.

An enraged bear, especially an adult, presents a formidable spectacle to its opponent. A typical threat may include a short charge, slapping of the ground with a paw, the utterance of a deep "huff," or an ominous chomping of the teeth. Any bear (or person) that chooses to disregard these clear warnings may be in line for a powerful, five-clawed swat or a cuffing-and-biting attack. Black bears are formidable animals; where they grow bold toward people, as around garbage dumps or in national parks, a casual attitude towards them can result in serious human injury.

# Grizzly Bear

**Although generally antisocial, grizzly and brown bears congregate around rich sources of food, such as salmon spawning grounds. It is under these circumstances that aggressive encounters (including occasional fights) are most frequently observed.**

The people who busy themselves with classifying animals can themselves be loosely classified into two types, the "splitters" and the "lumpers." Early in this century, when the splitters held sway, biologists fancied they could distinguish amongst eighty-seven species of grizzly and brown bears in western North America. This led to an absurd situation in which neighboring families of bears were probably assigned to different species. In more recent years, however, scientists have concluded that there is really only one, circumpolar species, *Ursus arctos,* which includes the brown bears from Europe, Asia, and North America, together with our grizzly bears. For the sake of convenience and pending further study, all the grizzlies and brown bears in mainland North America are often grouped in a single subspecies, *Ursus arctos horribilis.*

The *horribilis* in the scientific name refers

to the bears' reputation for size and ferocity, a reputation that is somewhat exaggerated on both counts. Grizzlies are not the hulking monsters of popular folklore. The largest animals are found among the brown bears that live on or near the coasts of British Columbia and southern Alaska and exploit the rich fishery of the salmon-spawning grounds. On this high-protein diet, individuals sometimes grow to 550 kilograms or more, but such giants are unusual. An average male "brownie" only reaches half that size, about the same as a typical grizzly boar from the Rocky Mountain parks. The females in all areas are somewhat smaller than their male counterparts. In the Yukon, for example, where forage is limited and the grizzlies in general are undersized, the sows seldom weigh more than 100 kilograms. It must be admitted, however, that this doesn't make them small by human standards. Drawn up on her hindlegs to get a better view, such a northern female would peer down on an intruder from the commanding height of two meters.

Grizzly bear

For animals of such heroic proportions, grizzly bears have remarkably humble tastes. In most places, they rely heavily on plant foods such as catkins, grasses, berries, and fleshy roots. Their principal prey (apart from fish, where they're easy to catch) are various rodents, from marmots down to mice. On the tundra, for example, grizzlies roll back the turf on boggy slopes to get at the lemmings that live underneath. In the mountains, the bears dig for ground squirrels, first shoveling out boulders and dirt with swoops of their powerful forepaws, then wheeling to pursue any squirrels that try to dodge away and hide. Sometimes, by reflex, the bears thunder off after stones that they have themselves dislodged. What with one thing and another, their success rate is usually not high.

Preying on large ungulates can be more toilsome still. In killing elk, for example, one of the grizzly's strategies is to harry the herd until a vulnerable animal can be singled out. Even when two or three bears cooperate (a rare occurrence in this largely asocial species), this may involve a run over several kilometers, with intermittent bursts of speed to close on the intended victim or cut it out of the herd. The prey is then attacked, dragged down, and bitten and shaken to death. Grizzlies can kill adult moose, bison, caribou, and mountain sheep as well, but they seldom take an ungulate in its prime, preferring to concentrate their attentions on young or ailing animals. Better yet, they like to acquire their meat as carrion. The larger the bear, the more likely it is to be capable of seizing or defending a carcass and the less likely it is to have to hunt for itself.

Thus, over most of their range, the supposedly bloodthirsty grizzlies function basically as herbivores and scavengers. Why then, if they are not driven on by a lust for fresh meat, do the bears attack and kill people in the mountain national parks? The first point to be made is that such incidents are exceedingly rare. Although the number of attacks is rising as the parks come in for increasing human use, the rate still stands at one injury for every 2,000,000 or so park visitors. By far the majority of documented cases involve sows with cubs. Unlike the forest-adapted black bears, which can shoo their youngsters to safety in trees, grizzlies spend much of their time in open habitat such as willow bars, berry patches, and avalanche slopes. Not surprisingly, they are not good at climbing trees. Their preferred defense is to avoid trouble, but if danger comes too close, a mother grizzly will take action to defend her family. This applies even if the threat comes from a large grizzly boar, since the male bears are known to kill cubs.

There are a number of things a person can do to avoid encounters with mother grizzlies. The ultimate safeguard is to stay out of the backcountry, because that is where most attacks occur. If you do go hiking, carry bells or clanking pots: given sufficient warning, grizzlies almost always flee. Sleep inside your tent, and avoid refuse-strewn campsites, since garbage-fed bears get used to human odor and lose their wariness. It goes without saying that it is wise to take extreme precautions with your own food and trash.

In the Yukon, grizzly sows have an extremely low reproductive rate — one cub every three years. Decreases in the birth rate, through the destruction of denning sites, or increases in mortality, through overhunting, may therefore be of special consequence.

You can also make a habit of watching for fresh bear sign. Grizzlies scrape shallow depressions in thickets, under windfalls, or beside streams in which to settle for their midday naps. Their tracks, which measure about twelve centimeters wide by twenty-five long, look a little as if they had been made by a huge person with one normal foot and one very badly fallen arch. The grizzly's front claws are long, extending well beyond the end of the toes, and this distinguishes its tracks from those of the black bear.

But what if, despite all your precautions, you surprise a grizzly at close range? Although there is no infallible defense, the best advice seems to be to stand quietly or to lie still with your legs drawn up next to your stomach and your arms over the back of your neck. Climbing a tree is a good idea if you can get out of reach quickly and if the bear isn't one of those eccentrics that will clamber up after you. Take what comfort you can from the knowledge that if the bear charges, it is likely to stop short of actual contact. By running, shrieking, or fighting back, you may provoke an all-out attack.

Historically, grizzlies have been the overall losers in their confrontation with mankind. Once "king of the beasts" throughout much of western North America, they have quite recently been exterminated over vast tracts of their former range. (The Swan Hills grizzlies of north-central Alberta, for example, are thought by some to be the last surviving remnant of the plains grizzly, and this population is itself now threatened.) The bears are exceptionally sensitive to human predation because of their low reproductive rate. The females do not reach sexual maturity until their fifth year at the earliest. Litters are small, usually one or two cubs, and the youngsters stay with their mother for two, three, sometimes four years. Only after she abandons her youngsters does the female breed again. Over the millenia, this rate of increase was sufficient to sustain the population, but that was before the introduction of firearms.

Grizzlies are also vulnerable to the disturbance of their habitat, by blatant or subtle means. Biologists speculate, for example, that with increasing air traffic in the North, low-flying planes may place an intolerable stress on arctic grizzly bears. There is also concern that the construction of northern pipelines will intrude on areas that the grizzlies favor for their winter dens. Typically, the bears dig their burrows on isolated, shrubby hillsides where the snow piles up in deep, insulating drifts. They remain underground from about October to April or May; the cubs are born in the dens in midwinter. Some biologists believe that the availability of suitable denning sites already places an upper limit on the numbers of northern grizzlies. If this is the case, fewer den sites can only mean fewer grizzly bears.

# Polar Bear

Polar bear
---International boundaries

In the winter, the temporarily ice-bound expanse of Hudson Bay and certain other arctic waters appear to furnish little in the way of a livelihood, but the polar bear, *Ursus maritimus,* is well equipped to thrive in these seeming wastelands. It possesses long, dense fur and a layer of fat for insulation, white coloration for camouflage, and the prominent canines and small, jagged cheek teeth that are required for dealing with its main food, the ringed seal. (This dentition is a reversion from the slightly more omnivorous form that is displayed by other bears.)

The ringed seals dwell in an other-worldly realm beneath the ice, but they are brought to the surface, and into contact with the polar bear, by their requirement for fresh air. Their lifelines are the breathing holes that they maintain in the cracks that accompany pressure ridges and in temporary channels formed where the ice separates. Beneath the snow that may drift over these breathing holes, seals often excavate hauling-out dens and birth lairs, which the polar bear seeks out with its keen sense of smell. Upon locating a seal's hideaway, a bear typically pauses a few meters away, listening until it hears the seal arriving in its cozy retreat. All of a sudden, the bear catapults forward and, with its powerful forelegs, bursts through the thirty centimeters or more of compacted snow and ice that form the ceiling of the chamber. This hunting method is most successful in April and May when the bears fatten rapidly by digging the highly vulnerable seal pups out of their birth lairs.

In the summer, when the ice is breaking up and breathing holes are exposed, a polar bear works them with patient cunning, often lying on its stomach for an hour or more, on the chance that a seal will make the mistake of surfacing nearby. When seals are discovered basking on the ice, the bear carefully stalks them, either by hiding amongst irregularities in the surface of the ice, while approaching in a crouch, or by swimming deviously through pools and channels until the seals are within range of a final rush. Even in open water, the ever-wary seals are not safe, for a polar bear

In the fall, before Hudson Bay is sufficiently frozen to permit sea travel, a polar bear prowls the coast on Cape Churchill. Its shape and size remind one of the grizzly, which has been known to interbreed with the polar bear in captivity.

may sneak up within striking distance by swimming rapidly when the seal's head is up, and floating motionless, like an ice floe, when the seal is submerged.

The winter and summer diet of the polar bear also includes bearded and harp seals and occasionally narwhals. A polar bear has even been known to take on a beluga five times its own weight, when the whale was trapped in a large, ice-bound pond. The bear probably lay in wait at the edge of the ice until the whale surfaced nearby, then, with a forepaw, delivered a mighty blow to the head. When sheer strength will not secure a meal, more subtle means will, as when the bear hunts seabirds by diving and swimming underwater, then surfacing, jaws agape, beneath the unsuspecting bird.

But it's the availability of a single species of prey — the ringed seal — that largely determines the movements and seasonal distribution of polar bears. In the southern Beaufort Sea, for example, polar bears travel great distances to follow the pack ice and the good seal hunting that it affords, drifting south to the mainland and east to Baffin Island through the summer, then moving seaward again with the new ice of winter. In Hudson Bay, pack ice is virtually absent between August and November, forcing the bears to seek refuge ashore on the coasts and islands.

Polar bears on the west coast of Hudson Bay (a population of 200 or 300) pass the late summer and autumn in a very leisurely fashion, using up some of their fat while consuming mainly grasses and marine algae. (They also eat garbage, where it is inadequately disposed of, and have become notorious problems around the town of Churchill.) To beat the heat and, to a lesser extent, the insects, these bears resort to a behavior not practiced by polar bears farther north: they dig summer dens, usually along eskers or in peat banks. While other animals are savoring the dwindling warmth of autumn, these hardy bears may be passing their most delightful hours by withdrawing to the darkness of deep burrows, to lie with their bellies pressed against the permafrost.

When autumn deepens into winter, dens are excavated in snowbanks for an opposite purpose — as shelter against the cold. In the confines of one of these one- or two-roomed hideaways, complete with ventilation shaft, sleeping alcoves, and a sealed entranceway, a pregnant female (rarely a male or a nonpregnant female) will pass five or six months in a lethargic state, though not in actual hibernation.

One of the largest and most concentrated maternity denning areas in the world is on Cape Churchill, in a relatively treeless area centered about fifty kilometers inland, between the Churchill and Nelson rivers. Here, the females break out of their icy dens between late February and early April, usually in the company of one or two cubs (already about three months old). After she has spent a couple of weeks in the vicinity of the den, feeding on snow-covered vegetation and watching her cubs exercising their new freedom, the female follows her urge to move her family seaward. It's a long, halting march, for the young cubs must rest and nurse several times a day. At each stop their mother solicitously constructs shallow beds and allows her youngsters to curl in the warmth of her fur. If a cub is especially cold and frightened, it may even complete part of the journey riding on its mother's back. Once on the sea ice, the cubs follow their mother closely as she prowls about in search of seal dens. The young begin their tuition in the art of seal hunting in their second spring, and by the time they are 2½ years old, they are usually on their own.

A major threat to polar bears' survival as individuals, and as a population, is mankind. With the world polar bear population at around 20,000, and with its hunting restricted by an international agreement, the bears would seem to be guaranteed a secure future. But there are new and unknown threats to their survival, such as oil pollution in arctic waters. Another ominous development is that polar bears, because they are at the top of arctic food webs, have been found to be concentrating pollutants such as polychlorinated biphenyls (PCBs) in their tissues.

# Raccoon

It was in the eighteenth century that the great Swedish biologist Karl von Linné modestly undertook to bestow a scientific name on every living thing. Inevitably, his ambition at times exceeded his knowledge, as in the case of the raccoon, which he described as *Ursus lotor,* "the bear that washes." In this he was wrong on two counts. In the first place, of course, the raccoon is not a bear; instead its closest kin are such exotic creatures as the coati, kinkajou, and panda, with which it shares the family name Procyonidae. In recognition of this relationship, the raccoon is now known as *Procyon lotor.*

Linnaeus's second misconception — that raccoons insist on washing their food — is still widely held, though careful observers have long since discarded it. What really happens is this: a raccoon that is kept in a cage, with no opportunity to catch its own prey, will sometimes drop bits of food into its water dish and dabble after them. Or it may try the same game with stones and chips of wood. One pet raccoon customarily ended its sessions of water play by urinating in the dish and then upsetting it. Whatever its objectives, they clearly had nothing to do with hygiene!

To understand the behavior of these captive animals, one has to study the habits of raccoons in the wild. This is not as simple as it

**Young raccoons generally leave their natal dens at six or eight weeks of age, by which time they can walk, run, and climb. But they can't keep up with their mother until they're about three months old and so are left behind when she goes to hunt.**

may sound, since they are largely nocturnal: they usually doze away the day in trees or in tramped-down marshland beds. Still, with the aid of a flashlight, observations can be made. The trick is to take up watch in the evening beside a body of water where the animals come to feed — someplace where you've previously noticed their five-toed, handlike tracks. Then, play your light over the shoreline and shallows until you pick up the orange or greenish glare of a pair of eyes. You will be able to follow the animal's actions without much fear that the spotlight will frighten it. Perhaps your raccoon will shuffle along the beach, pausing frequently to reach into the water and grope in the mud. Or it may wade out up to its snout in water and work the bottom with its front paws, poking, probing, feeling into the slimy crevices around rocks and logs. And all the while, it will likely stare off into the distance as if nothing were going on.

The object of these underwater searches is often crayfish, which the raccoon locates by touch. Unlike most other carnivores, whose tactile sense is concentrated in their muzzles, the 'coon has ultra-sensitive fingers, with powers of discrimination on a par with ours. Thus, it is not a "washer" but a "feeler," and its water play in captivity can best be viewed as a make-believe crawdad hunt.

Although raccoons apparently prefer aquatic prey when it's available, crayfish are by no means their only fare. Able both to swim and to climb, raccoons can reach almost any potential source of food, whether it's a lodge full of muskrat kits in the middle of a marsh or a clutch of owl eggs in a ten-meter tree. The world lies before them like a sumptuous buffet set with duck nests, raspberry bushes, earthworms, crabs, fruit trees, garbage, and sun-ripe corn. Although they eat a wide array of animal food, their diet in most places and seasons includes a high proportion of plant matter, usually fifty percent or more.

The raccoon's omnivorous diet is one reason that the species can thrive over such a large and varied portion of North America. Shortage of water and extremely long winters seem to be the only factors that limit their range. In recent decades, they have been expanding their sphere northward into the prairie provinces, most notably in Manitoba. During the 1950s, when raccoons first started to turn up around Winnipeg, people were so surprised to see them that several animals were solicitously "returned" to the local zoo. Since then, they have become relatively abundant around potholes, marshes, and wooded coulees in the south and have occasionally been sighted in the boreal zone as far north as Granville Lake. Although the evidence from Saskatchewan and Alberta is scanty, there are hints that a similar population expansion may be occurring there.

The raccoon's accommodation to conditions on the prairies is thorough-going. People used to think, for example, that the females required hollow trees in which to bear their young. But in areas like western Canada, where such den sites are rare, resourceful mother raccoons shelter their offspring in granaries, haylofts, and thickets or in burrows dug by woodchucks, foxes, and skunks.

What's more, the average prairie litter of five ratlike cubs is significantly larger than those born in the south, presumably, in part, as compensation for higher wintertime losses. In years when the snow cover lingers well into spring, more than half of the yearling raccoons have been known to die.

These deaths occur in spite of several weather-wise tactics that northern raccoons adopt against the cold. Although they do not hibernate, they make it through the winter by living off their fat. To this end, they gain weight prodigiously in the fall; then, with the first permanent snowfall of the season, they settle their pudgy bodies into the dead trees, brush piles, or snow dens in which they will stay for weeks or months at a time. Frequently a female and some or all of her litter snuggle into the same refuge. Family ties, which begin to weaken in midsummer as the cubs grow up, suddenly strengthen again at our latitude with winter's approach. The youngsters don't strike out on their own until the following spring, long after their southern age-mates are leading independent lives.

**A member of a family that is typically tropical and subtropical in distribution, the raccoon has made a remarkable accommodation to northern conditions.**

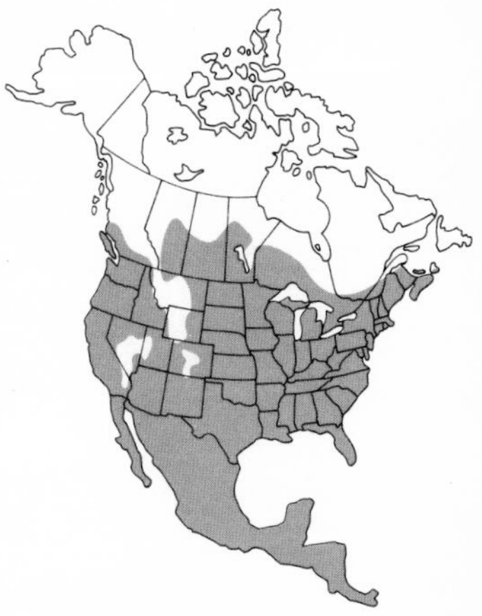

**Raccoon**

# Marten and Fisher

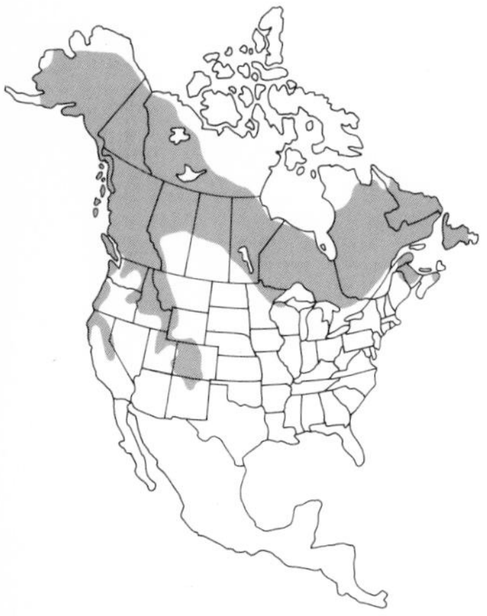

Marten

More than any other member of the weasel family, the marten, *Martes americana,* has become adapted for life in the forests. Its yellowish-orange to near-black fur blends in easily with the forest shadows. For climbing trees, nature has provided it with sharp, semiretractable claws; and for balancing, it has a long, bushy tail, about half the length of the rest of the body. The most arboreal of our carnivores, the marten is at home in mature mixed woods or coniferous forests containing stands of white spruce and jackpine.

Within these habitats, it sometimes shows a preference for areas where there are many deadfalls and tangles of brush. When hunting, it may charge into these clumps of vegetation or fallen timber with a sudden leap from the uphill side. A snowshoe hare or red squirrel may be flushed by this stratagem; failing that, the hunter can always nose about the nooks and crannies for a mouse or vole. In the winter, the marten may tunnel long distances under the snow, investigating possible sources of a meal. Although most of its hunting is done on the ground, it is also nimble-footed in the trees. Indeed, it makes a game of climbing a conifer or cottonwood and "skydiving" out of it from a height of several meters or more. Tracks in the snow have shown where a marten has landed ten or twelve meters from the trunk of the tree. Perhaps these aerobatics keep it in practice for pursuing red squirrels and flying squirrels in wild chases through the treetops.

Tree squirrels are usually not the marten's most important prey, however. Its main fare is more often mice and voles, particularly Gapper's red-backed vole, a common inhabitant of mature coniferous forests. Ground squirrels, pikas, snowshoe hares, and birds are also eaten. When hunting is good, a marten may procure more food than it immediately requires and cache a surplus kill in a hollow tree. Late summer and fall may find this opportunistic omnivore subsisting almost entirely on insects or berries. Around a good berry patch or a large supply of carrion, martens become temporarily sedentary, but usually they travel widely in search of food. Most active around dawn and dusk, they may cover several kilometers in a single night.

During the day, they rest up, sometimes close to a kill, in a sheltered spot such as a hollow stump, a rock pile, a ground burrow, or a crevice around fallen trees. Often a red squirrel is temporarily evicted from its nest to make way for an itinerant marten. If other lodgings are not available, the marten is quite prepared to spend the day sleeping high on a branch in a tree.

While most dens are used only for short periods of time, occupancy is more permanent when a female is raising her young. The kittens are born, most commonly three per litter, in late March or early April. Soon after the litter is weaned, at about six weeks of age, the mother breeds again, but the next litter doesn't arrive until the following year. (This exceptionally long gestation period is a result of "delayed implantation," a phenomenon that also occurs in most of the other mustelids and in bears. What happens is that, shortly after conception, the development of the embryo is temporarily arrested; it just sits in the female's reproductive tract, rather like a seed lying dormant in the ground, until the time comes for growth.) At about three months of age, the young martens reach their adult weight, approximately 1,100 grams for males, 750 grams for females. They don't breed until their third summer; consequently first litters are usually born to three-year-olds.

This relatively late age for reproduction, combined with a comparatively small litter size, causes marten populations to be slow to recover from a decline. In the absence of human interference, their numbers are fairly stable, limited only by the availability of food; occasional predators such as coyotes, fishers, golden eagles, great horned owls, and cougars have little impact. But because their curiosity

Although it spends most of its time on the ground, the marten takes to the trees with ease. The light-colored patch on the throat and breast is an aid to identification, but it is occasionally absent.

Crevices in rockslides sometimes serve as dens for the fisher. Other den sites are snow burrows, hollow logs, and brush piles; even the body cavity of a large carcass will serve, until the fisher eats itself out of house and home.

makes them very easy to trap, martens have in the past been nearly exterminated in many areas. Today, martens are rarely observed in the wild, partly due to their secretive habits, but also because of the continuing pressure that trapping places on the population. A further threat to their numbers is the loss of suitable habitat. After large tracts of coniferous forests are logged or burned, martens are unable to survive in the new community of deciduous growth that springs up.

A close relative of the marten, with similar habitat requirements, is the fisher, *Martes pennanti.* Not only are both these species confined largely to coniferous forests, they share many other basic similarities: in their choice of denning sites; in their ability to climb and jump from trees; in their solitary, primarily nocturnal habits; and in their reproductive biology (one exception being that female fishers first breed at one year of age instead of two, as in the marten). They are quite different, though, in their feeding habits. The fisher's larger size (about a meter in total length, including the tail) better suits it to hunting larger species of prey. Throughout most of its range, the main entrée in the fisher's menu is the snowshoe hare, which the fisher locates by a random search of thickets and windfalls, and captures by short, lightning-quick chases.

The fisher also preys heavily on porcupines and is one of the few carnivores to have mastered the art of killing this well-armed rodent with impunity. As long as the porcupine is in a tree or in a den with only one entrance, the fisher has difficulty in getting past the prickly posterior of its intended victim. The fisher's strategy, therefore, is to try to surprise its quarry on the ground during one of the porcupine's infrequent trips between its den and a feeding tree; to this end, the fisher regularly inspects porcupine dens, often on purposeful, round-trip tours of its home range. When a porcupine is discovered in the open, the fisher immediately launches an attack on the unarmed and vulnerable face of its prey. In defense, the porcupine attempts to present its backside or bury its nose against the base of a tree, sometimes charging backwards, thrashing its tail. But the infinitely more agile fisher presses the facial attack until the porcupine succumbs. The fisher's expertise is also exercised in devouring its prey — it tears open the quill-free abdomen and skins the hide back as it feeds. The fisher may pick up a few barbs in the whole process, even ingesting some of them, but strangely enough, the quills seldom cause any apparent harm.

Although porcupines and snowshoe hares are usually the most common food items, fishers find many other sources for a meal, including deer and moose as carrion, muskrats, red squirrels, northern flying squirrels, mice, shrews, and even berries and nuts in season. But one of the few foods that fishers rarely, if ever, eat is — you guessed it — fish.

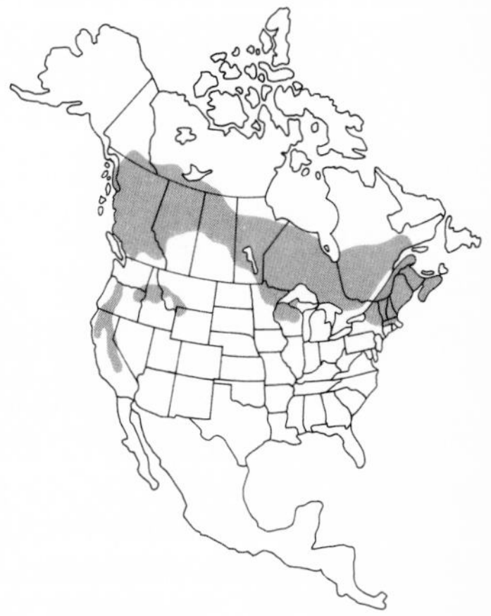

Fisher

# Weasels

The weasels are a very numerous group of carnivores that make up in ferocity for what they lack in size. The two larger species, the ermine, *Mustela erminea,* and the long-tailed weasel, *Mustela frenata,* are confusingly similar in appearance, both of them showing tawny summer coats, with light-colored chins and bellies, and black pencils on the tips of the tails. The summer coats do, however, show some discernible differences — for example, the underparts are white in the ermine and are generally light yellow in the long-tailed weasel. For winter, though, they adopt identical wardrobes. The shorter day lengths of fall stimulate a gradual moulting to new coats that are entirely white except for the black tail tips. In

In the winter, the long-tailed weasel, shown here, is a dead ringer for the ermine.

spring, another moult restores the summer coats. (Exceptions are found in areas where snow cover is uncommon: for example, the ermine usually wears its summer colors the year around on Vancouver Island, in the southwest corner of mainland British Columbia, and in portions of the northwestern United States.) Size is of little value in distinguishing between the two species: the overall length of body and tail is similar, averaging about twenty-five centimeters in the ermine, thirty-five in the long-tailed weasel. For field identification in winter, a more reliable characteristic is the tail length: where the ranges of the two species overlap, the ermine's tail is less than three-tenths of the animal's total length, the long-tailed weasel's more than three-tenths.

We have a third species as well, one that is widespread but rare throughout most of its range. This is the least weasel, *Mustela nivalis,* the most diminutive of all the carnivores, achieving a maximum length of only twenty-five centimeters, one-fifth of which is its tail. Season by season, its coloration is very similar to that of the ermine, but its small size and the lack of a black tip on the tail are its distinguishing marks.

Weasels are diabolically efficient killers, extremely formidable in their own little worlds. The long-tailed weasel, for example, frequently preys on adult snowshoe hares, a feat that, if it occurred on a larger scale, would be equivalent to a house cat killing an animal the size of a pronghorn. Often, the weasel's technique is to pursue the hare relentlessly by scent, for as long as an hour if necessary, seeking an opportunity to sink its dagger-like canines into some portion of the hare's anatomy. The hare may repel the initial assault, but loss of blood weakens it and also provides an excellent scent trail for the hunter. It isn't long before the final encounter takes place. Then the scream of the hare may disturb the night, as the weasel's teeth, directed with the precision of a surgeon's scalpel, penetrate the back of the hare's skull or dislocate a neck vertebra. Hares and rabbits are especially vulnerable in their burrows or when young — even the least weasel is an occasional predator of young cottontails.

For the most part, however, weasels prey

Facing page: The yellow belly identifies this sinuous hunter as a long-tailed weasel in its summer coat.

After a fruitless inspection of a ground squirrel burrow, a long-tailed weasel pokes its head out. Its winter garb is starkly inappropriate when fall snows are late in coming.




on small rodents. The short legs of weasels and their long, almost snakelike builds ideally suit them for pursuing rodents along runways and into underground tunnels. The ermine and the least weasel both specialize in capturing meadow voles or, in the far north, lemmings. In many areas the cycles of abundance of these rodents are closely followed by fluctuations in the weasel population, revealing the extent of their dependence on a single species of prey and suggesting, to some biologists, that the interaction between predator and prey actually causes the cycles. Ermine frequently take other prey, such as mice and shrews, but the least weasel finds shrews distasteful.

The long-tailed weasel, while it consumes plenty of mice and voles, avoids the booms and busts in the economy of its near relatives by being more flexible in its diet. When voles are in short supply, it turns more to hares, rabbits, pocket gophers, shrews, and duck eggs. It even pursues and catches chipmunks in the treetops, and also frequently preys upon ground squirrels, which may be caught above ground and killed in the typical weasel fashion, that is, by hugging the victim's back, raking with the hind feet, and delivering a well-placed bite to the back of the neck or head. Below ground, the long-tailed weasel employs an easier and safer technique: the ground squirrel is quickly dispatched with a suffocating bite to the throat. When other food is scarce, the long-tailed weasel may turn to marauding hen houses, slaying as many as a dozen chickens in a single night's work. But weasels generally compensate farmers for these misdemeanors by regularly disposing of enormous numbers of mice and rats.

Excessive killing, such as sometimes occurs in a hen house, also occurs in the wild, especially in the fall, and is characteristic of all three species of weasels. This isn't simply a wasteful lust for slaughter, however, since the weasels take the surplus bodies underground or bury them for future use. Often, but not always, the storeroom is located in the weasel's den — a hollow log or the usurped burrow of a mouse or gopher. The den also contains a cozy nest of grasses and plant fibers, lined with fur or feathers plucked from prey. Temporary homes are sometimes made under the snow in the winter nests of lemmings or voles. Unwelcome guests, the weasels devour their hosts, and often the neighbors as well, then move on in a few days or weeks to find new lodgings.

A more permanent nest is used for raising the young, usually four to eight per litter, which are born in the spring. Young weasels mature rapidly, the females before the males, and disperse by fall. At maturity, the males are considerably larger than the females, a circumstance that allows them to exploit slightly different food resources and hence reduces competition between the sexes. Devices like this may be particularly important to long, thin animals like weasels, because they readily lose body heat to their surroundings and so have very high food requirements. Other consequences of their shape are found in their winter activities. In order to satisfy their ravenous appetites, long-tailed weasels make nocturnal hunting trips of up to two kilometers or more. In the morning, the record traced in the snow reveals that the weasel has nosed into every hole or likely looking cranny in its path. The smaller weasels, being more susceptible to heat loss, tend to adopt a stay-at-home strategy, confining their activities to a smaller area, while exploiting the warmer environment beneath the snow. Ermine, for example, remain under the snow blanket whenever the outside temperature is below minus thirteen degrees Celsius. The long-tailed weasel is too large to forage in such cramped spaces.

Staying "indoors" also helps weasels to avoid predators, such as the red fox, which kills many weasels, even though it is reluctant to eat them. Hawks and owls are also major enemies, but the conflict is not entirely one-sided. Long-tailed weasels have been observed harrying these birds from fence post to fence post, apparently in an attempt to induce their avian foes to drop some recently caught morsel. And the long-tailed weasel's belligerence is not confined to birds. There are several instances of people being attacked after trapping and releasing these weasels or otherwise provoking them.

# Mink

Very similar to the weasels, both in appearance and habits, is the mink, *Mustela vison.* The dark color of its sleek fur is interrupted by the white chin patch, characteristic of the genus, and often by dashes of white on the throat or belly as well. Its serpentine form might be mistaken for that of a weasel were it not for the mink's greater weight and heavier musculature. Males weigh about two kilograms, sometimes as much as four, while females, at about one kilogram, still dwarf the largest of the weasels.

In the water, the mink reveals its kinship with another member of the family, the otter, by propelling itself easily with the aid of its partially webbed hind feet. The diet of the mink reflects its amphibious capabilities. On the west coast, it feeds mainly upon small crustaceans and fish, caught in the intertidal zone. Inland, it may insinuate its lithe body into the thickets along a water course, where it can make its meal on meadow voles, cottontails, frogs, snakes, or insects. In marshy areas, ducks, coots, and their eggs are frequent fare. Muskrats are a favorite food but cannot be counted on as a staple: a healthy, mature muskrat can often defend itself if it stands its ground or backs itself into a hole. Occasionally an epidemic sweeps through the muskrat population and the carrion is scavenged by mink. At other times, they may prey on those luckless muskrats, especially the young, that are forced to wander in the open, perhaps by intolerant neighbors, perhaps by a shortage of food brought on by drought or exceptionally deep ice. Even then, it is usually only the male mink that is large enough to take on this weighty rodent.

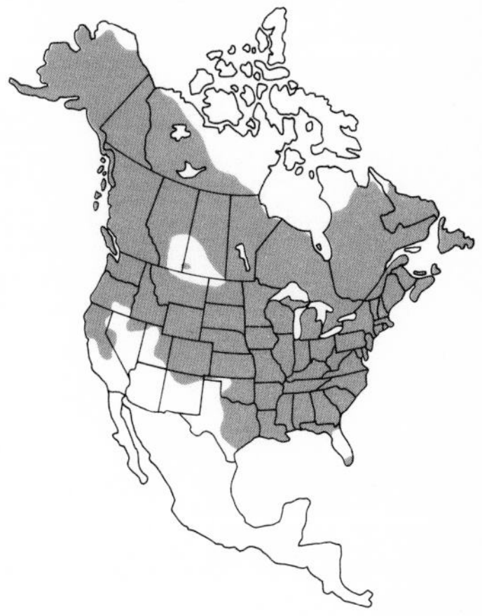

Mink

The mink is adept at fishing. One common tactic on the coast is to herd a school of small fish into shallow water, then snap up any that happen to become beached momentarily. The mink may also dive for fish, especially in landlocked pools, but it is imperfectly adapted

to search for them underwater. Its eye muscles cannot fully compensate for the loss of refractive power when the cornea is submerged, so underwater images remain out of focus. Perhaps this is why the mink sometimes adopts the strategy of a patient angler, waiting on a vantage point, such as the top of a flat rock, for a fish to approach, and then plunging suddenly after it. If the fish is a bit too slow, the mink will grab it in its teeth, aided by a lightning quick maneuver of its sinuous neck. One observer watched a mink make seven successful catches by this method, depositing each small fish in the center of a rock before administering a death-bite to the back of the head. The fish were then carefully carried, one at a time, across the stream to be deposited in the den. This food-storing habit has been observed on other occasions as well — one den was found to contain nine muskrats, four ducks, and five coots, a sizable feast for a single mink!

Most dens are located close to water, most often in a muskrat burrow whose original occupant has either been devoured or has chosen to vacate. Typically there are one or two entrances beneath or just above the water's surface; from here the tunnels slope upwards, perhaps undermining the roots of a tree, perhaps joining with a network of subterranean passages that have several surface entrances. Dens may also be found in hollow logs or abandoned beaver lodges.

In the winter, the female occupies one or two such dens, confining her activities to the surrounding area, possibly twenty hectares or less in extent. The males travel more widely as they investigate one den after another. Any time of the year, most mink shun the day, and during the coldest winter weather, they don't venture forth at all but remain in their burrows or prowl along creek or river banks in the extensive air spaces that occasionally form beneath the ice. In coastal areas, however, the activities of mink are quite different. Their lives are regulated by the tides, since foraging is best at low tide, whether this occurs at night or during the day.

The wanderings of the males bring the sexes together briefly for breeding, sometime between late February and early April in inland areas. By early May (or later on the coast), litters of two to ten kits have been born. Blind and deaf at first, and only thinly covered with fine white hair, they develop rapidly. When only a few weeks old, they are able to travel with the mother from one den to another or accompany her on hunting expeditions.

As youngsters, mink love to play at a variety of strenuous games such as stalking and pouncing upon an unsuspecting littermate. Amidst much hissing, squealing, and growling, the attacker attempts to bite the other's abdomen and give its pretend victim a good shake, while the defender lies on its back and tries to protect itself with its feet. Sometimes they gallop after one another, biting and wrestling, or plunge into the water and roll about. As adults, they retain some of this playfulness: borrowing a trick from the river otter, they occasionally toboggan down snowy slopes on their bellies. By the middle or end of their first summer, the families break up, and before they are a year old, the young mink are fully grown and ready to reproduce.

**Marshes provide ideal habitat for mink, as do stream and river banks, lake shores, and tidal zones.**

# Wolverine

Of all North American mammals, one of the rarest and most poorly understood is the wolverine, *Gulo gulo*. In the absence of fact, a great deal of fiction has built up around this elusive denizen of the tundra and boreal forest. One of the most outlandish of these tall tales claims that the wolverine devours so much when feeding at a carcass that it must seek out two closely spaced trees and squeeze its overstuffed belly between them to relieve itself of excessive flatulence, before returning to finish off the carcass. Another tale, which seems nearly as improbable, actually turns out to be true: the wolverine does on occasion sit on its haunches and gaze into the distance, holding up a forepaw to shield its eyes from the sun, just as a person would do. The same mixture of fact and fiction shows up in some of the colorful names that have been applied to the wolverine. "Skunk-bear" is apt, since the animal rivals the skunk in possessing a pair of anal glands capable of ejecting a malodorous yellowish-green musk to a distance of three meters. "Indian-devil" originated with trappers who, marveling at the boldness and apparent maliciousness with which the wolverine stole trapped animals and bait from trap lines, thought it the work of the devil incarnate. In the Old World, where the tale about the two trees originated, the wolverine is unfairly maligned with the name of "glutton."

Wolverine

Although its appetite is not really excessive, no one has ever accused it of being a fussy eater. It consumes anything from wasps to whales, the former as larvae, which are plundered from wasps' nests, the latter as carrion, washed up on beaches. Always an opportunist, the wolverine's summer diet includes birds' eggs and fledglings, fish, frogs, mice, voles, squirrels, marmots, snowshoe hares, beavers, and marten. It doesn't turn down vegetarian food either and may live entirely on berries for a period in the fall. In the winter, it is primarily a scavenger of carrion. It is particularly well suited to this niche by having extremely strong jaw muscles and massive carnassial teeth; thus equipped, the wolverine can chew frozen meat or crush bones to extract the last bit of nourishment from an old carcass.

This scavenging mode of existence explains why wolverines are so scarce: each adult must support itself on a food supply that is scattered over vast areas, generally hundreds of square kilometers in extent. Traveling widely, the wolverine scouts its home range for a meal, its acute sense of smell ever alert as it lopes along in a powerful, effortless gait. Every few hours it may stop to rest and sleep; sometimes it pauses in its travels for a playful roll in the snow, but in good weather its need to travel urges it on.

While its winter food is primarily carrion, the wolverine is also an accomplished big-game predator in its own right, occasionally killing mountain goats, deer, elk, muskox, or caribou. Not possessing the speed of the wolf nor the stealth of the cougar, the wolverine pursues its prey by a direct chase, comparatively slowly but with great endurance. On firm ground, the attempt is almost never successful, but in late winter, the wolverine's widespread toes give it the advantage in traversing deep snow. Under these conditions, the fury of a fifteen-kilogram wolverine may conquer even a bull moose, an animal thirty times heavier. The wolverine's method of attack, for the moose and other cloven-hoofed mammals, is to leap onto the victim's back from a tree, if the opportunity presents itself, or from the ground. Clinging tenaciously, it tears a hole in the prey's spine or severs the jugular vein. When a large amount of food is obtained, most of the booty is cached for future use, pieces being buried in the ground or the snow, lodged in trees, or covered with rocks.

Caches are set up the year around but have a special significance in the spring when the females use them to supply their pups with their first solid food. The pups are born any time from January through April, two to four per litter. Shelter is provided by a den, which is located in such sites as a rock slide, under the roots of a windfall, or perhaps in an old beaver lodge; often the den is little more than a long tunnel in the snow. By the time the young are six to eight weeks old, the mother is regurgitating food for them, usually from one of her caches. The den is abandoned at about this time, but the pups are still not strong enough to accompany the mother on her longer hunting expeditions. The mother breeds again in midsummer and by the time her new family arrives, during the following spring, the previous litter has dispersed.

The wolverine's habit of setting up caches extends to other items besides prey or carrion, contributing to its bad reputation with trappers and people who dwell in remote northern areas. Enticed by the scent of food, wolverines may break into cabins or caches of supplies, carrying off and hiding food, snowshoes, skis, traps, and just about anything else.

Like so many other species, the wolverine has lost out in the conflict of interests between humans and wild animals. Considered a destructive nuisance by trappers and possessing a unique fur that is highly prized by northerners as a frost-resistant trim for parka hoods, the wolverine, rare though it is, becomes rarer still through the efforts of trappers. It is, however, usually clever enough to avoid getting trapped, and even when caught, it will, like some other animals, chew off toes or a limb to escape. The distance that it can drag a trap before cutting itself free, ten kilometers and more, testifies not only to the terrible cruelty of the leg-hold trap, but to the endurance of the wolverine.

Other than people, wolverines have few enemies — the occasional cougar or a porcupine whose quills may exact a slow vengeance after their owner has been eaten. The only natural predator of any real significance is the wolf. Where a wolverine can find a tree to climb, it is safe, but caught in the open, it is vulnerable to an attack by a wolf pack. However, the wolf is more important to the wolverine as a provider of carrion than as an enemy, so much so that the decline in wolverine numbers is likely due in part to the reduction in wolf populations occasioned by wolf-control programs. Never very abundant, the wolverine is now all but extinct in the aspen parkland and is found mainly in British Columbia and the North.

**One of the largest members of the weasel family, the wolverine has a well-deserved reputation for strength and ferocity.**

# Badger

To borrow a phrase from a writer of an earlier day, the badger, *Taxidea taxus,* looks like "a small Bear that has been flattened somehow." And so it does, with its broad head; long, curved front claws; and pigeon-toed, waddling gait. Although, at a weight of five or ten kilograms, it can hardly be expected to share fully in the legendary strength of the bear, it is nonetheless powerful, as befits an animal that has to labor for its livelihood.

The badger is a living steam shovel. Its principal prey are large burrowing rodents, such as the thirteen-lined and Franklin's ground squirrels, with which it shares the native prairie and the aspen parkland, respectively. In mountainous areas, it preys on such species as Columbian ground squirrels. On first consideration, it seems improbable that any animal could claw its way along a rodent tunnel, enlarging the passageway as it goes and spraying out dirt behind, and yet make faster progress than the occupants, which supposedly have nothing to do except frisk away through some other door. But perhaps squirrels often

Squat and powerful, the badger can dig itself out of sight in a minute or two by loosening the soil with its front claws and sending it flying with its hind legs.

**Although burrowing rodents are the badger's principal prey, it consumes a varied diet that may, on occasion, include rattlesnake!**

get trapped in underground cul-de-sacs. Or perhaps they are sometimes taken unawares, since badgers usually make their raids at night, when the squirrels are at rest.

Squirrels frequently do escape, of course, and badgers have been known to take the precaution of plugging alternate burrow exits before ripping in along the main passageway, although this sophistication is by no means commonplace. The occasional badger may also dig itself right inside a squirrel tunnel, partially block off the enlarged entrance from within, and then lie in ambush beneath another, untouched doorway, ready to gobble up any of the residents that happen into its waiting jaws. From midsummer on, as more and more of the ground squirrels settle into a prehibernation stupor, such cunning must become unnecessary.

Even when wide awake, ground squirrels don't demonstrate much savvy about the badger's ways. They will continue to use burrow systems that have been partially dug out, in spite of the badger's habit of reworking the same digs time and time again. Another group with a shortened life expectancy must surely be those cottontail rabbits that take shelter in badger holes. Badgers also prey on pocket gophers, mice, voles, ground-nesting birds, poultry, birds' eggs, and insects, especially bumblebees, which they consume nest and all. Surprisingly, they can swim and catch fish as well. Surplus food is buried underground, presumably for future use.

**Badger**

Surely the most interesting aspect of the badgers' food-getting behavior is their willingness to hunt alongside coyotes in certain rare instances. These unusual relationships are evidently based on mutual respect. A group of coyotes will, very occasionally, kill a badger, and badgers sometimes prey on coyote pups, but in single combat between adults, the two species are evenly matched. Their hunting partnerships may appear one-sided, with the badger doing all the heavy work of digging for rodents, while the coyote stands by to snatch up escapees. But there must be benefits for the badger, too. Not only does it tolerate this arrangement, which it needn't do, but it may purposely follow its canine accomplice, relying on the latter to stop now and then so it can catch up. Sometimes, these unlikely confederates will make a joint rush through a rodent colony, each taking advantage of the scurrying confusion caused by the other's attack. Or the badger may dive underground to pursue an animal that the coyote has chased into a hole.

These interspecific partnerships are all the more remarkable since badgers rarely associate with members of their own species. An adult male may share parts of his home range with two or three females, but the animals do not consort except during the mating season, which falls in July and August. The young, usually two to a litter, are not born until the following spring. Like martens and several other members of the weasel family, badgers exhibit "delayed implantation," which retards the development of the embryo. This mechanism is particularly important for badgers at our latitude, because they become torpid in late winter, when breeding would otherwise have to occur.

The young are born in a special natal den, one or two meters deep, that the mother excavates. Unlike ordinary summertime burrows, which are seldom occupied for more than a single day, this one may be inhabited for three consecutive weeks. It can be identified by the extra-large entrance mound of soil, into which quantities of badger hair have become mixed. In western Canada, the best time to look for such a habitation is between mid-April and early June. With luck, you might see the youngsters gamboling around the doorway or their mother taking a nap in the morning sun. Such sightings must once have been common, before intensive agriculture caused badger populations to decline throughout much of the species' range.

# Skunks

Striped and spotted skunks, *Mephitis mephitis* and *Spilogale gracilis,* are the chemical warfare experts among North American mammals. Unlike most carnivores, whose ultimate defense lies in the *sauve qui peut* of tooth and claw, the skunk can repel its enemies from a safe distance by spraying them with a nauseous, eye-stinging musk. This fluid, which is produced in two glands at the base of the skunk's tail, is usually reserved for emergencies. Skunks are peaceable creatures and, given the chance, they will lollop away from danger or bluff their way out with threats.

In addition to the skunk, there are several species of mammals elsewhere in the world that use a stink-bomb in self-defense, and all of them are streaked with black and white. Evidently, these distinctive markings are a warning to would-be predators, a conspicuous and unmistakable reminder of their bearers' special powers. Skunks emphasize this message in their defense displays. The agile western spotted skunk, which occurs in southwestern British Columbia and most of the western states, flings itself up into a handstand and walks forward on its front feet to advertize its capabilities. It can fire from this position as well. The striped skunk is less flamboyant.

The striped skunk, with the distinctive white bar on its forehead and the bold V on its back, is about the size of a domestic cat. By making it easy to recognize, the skunk's markings help to protect it from attack and permit it to make the most of its scent-spray defense.

The western spotted skunk, shown here, can be distinguished at a glance from its larger relative, the striped skunk. Although the general pattern of dots and streaks is the same on all spotted skunks, each invididual's markings are, to some degree, unique.

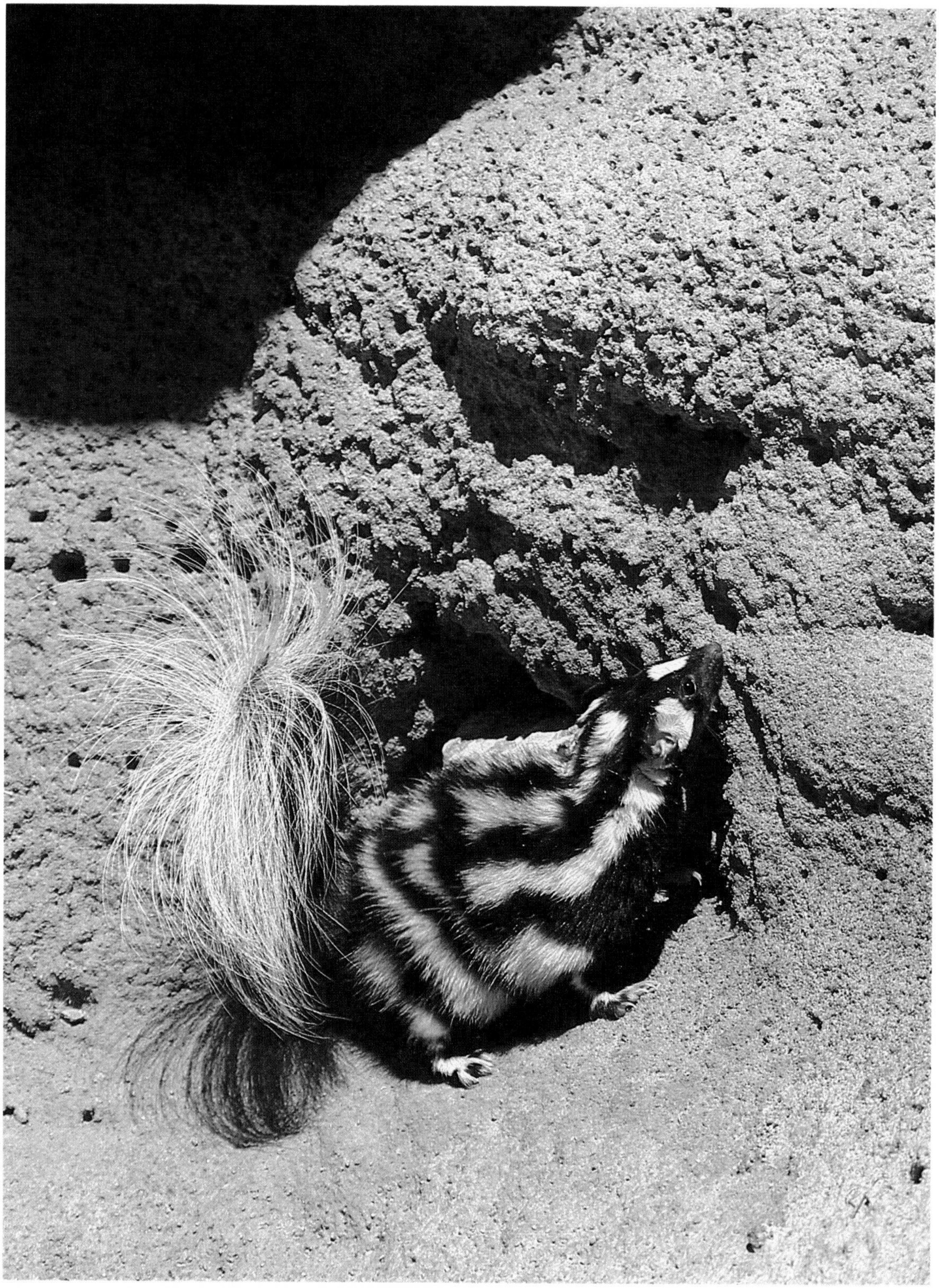

When cornered or closely pursued, it arches its back, plumes its tail, and meets danger face to face. Then, with its front feet, it drums out an ominous tattoo. As a last resort, it aims its backside at the target and lets fly with its reeking jet. By turning slightly as it shoots, a striped skunk can cover an arc of about forty-five degrees to a distance of three meters or more. This defense is devastatingly effective against all mammalian predators, from grizzly bears on down. Tasty and delicate though skunks are said to be, they are pretty much left to the great horned owl and other large raptors, which have almost no sense of smell.

There is no evidence that skunks purposely befoul one another. In fact, among striped skunks, the adults have little contact of any kind from late spring until early fall. From May until July or August, the females are preoccupied with their families, which average six or seven twittering kits. Their first home is often an underground den, perhaps an old woodchuck or badger burrow, perhaps the work of mother skunk herself. In preparation for the birth of her litter, the female may temporarily remove all nesting material, presumably so none of the helpless new arrivals will get caught in the tangle of dried grass. The youngsters don't need this precaution for long, however, as they gain strength quickly; their weight jumps by six-fold in as many weeks. By the age of two months, their infancy is over, the family breaks up, and each animal goes its placid way, dozing by day in croplands, fence rows, or shelters and hunting at night in its favorite meadows and fields.

Grasshoppers, grubs, and beetles are among the striped skunk's staple foods, but it also likes to snack on honey bees and will scratch at the hives to lure the insects out. A few sharp beats with the forepaws and down goes a bee, stinger and all. Caterpillars, on the other hand, are carefully prepared for consumption by rolling them on the ground to remove the unappetizing hairs. Birds' eggs are usually broken with the teeth; failing that they may be grasped in the forepaws and rolled back between the hind legs until they crack on something hard. Small mammals, amphibians, and plants round out the skunk's warm-weather fare.

Then along comes winter, to revolutionize this easy, itinerant life. Except for young litters with their mothers, which are more or less sedentary, few striped skunks use the same retreat for longer than a week during the summertime, but now, in October or November, they must settle down in one spot — under a building, for example, or in a ground burrow — to wait the winter out. In western Canada, striped skunks commonly stay in their winter dens for four or five consecutive months, emerging only briefly if there should be a thaw. Although they don't hibernate, their body temperature does drop slightly. Like bears, they become drowsy, lose their appetites, and subside into a lethargy. Female skunks usually hole up together, often in the company of a single male; on the average, half a dozen animals are found in these communes, huddled in a companionable heap for greater warmth. Yet, in spite of their energy-saving tactics, striped skunks lose about half their predenning weight before emerging in March. Not surprisingly, some of them starve, particularly females and juveniles that were underweight in the fall. Females are also particularly vulnerable to rabies (a disease to which striped skunks are notoriously susceptible). What with one thing and another, striped skunks seldom survive longer than three years, about a third of their potential life span.

**Striped skunk**

# River Otter

River otter

Consider the following scene: it is dawn, and you are gazing across the surface of a lake, still mysterious with early-morning haze, when a sleek, serpentine form arches quietly out of the water and then sinks from sight. A few feet back a second shape undulates into view; behind that, perhaps a third! A whiskered knob of a head pops up atop a thick neck, and the creature peers about. What is it? Not a sea serpent, as you might at first assume. Almost any clean, forest-fringed body of water in Canada or the United States provides potential habitat for the river otter, *Lutra canadensis*. A troupe of these lithe animals, traveling single-file as they often do, might easily give rise to rumors of a "sea monster."

In the water, river otters are as supple as seals and just as frolicsome, swimming now on their bellies, now on their backs, twisting, diving, and performing corkscrew turns. When it comes to actual travel, the animals resort to one of two basic strokes. If they're not in a hurry, they may paddle along with their short, powerful legs, using their heavy, tapering tails as rudders and spreading their toes to take advantage of the webbing on all four feet. To cover any distance, they will tuck their legs back out of the way and propel themselves by flexing their entire bodies up and down, beginning with the head and shoulders and continuing to the tip of the tail. In this fashion, otters can reach a top speed of ten or eleven kilometers per hour (about as fast as a canoeist can paddle), traveling either on the surface or underwater. Since they can hold their breath for over four minutes, they are able to cover half a kilometer or more without coming up for air, an ability that can be particularly useful in winter, when they often swim under the ice.

River otters have a reputation for being great travelers. Except for the breeding females, which settle down temporarily to raise their young, these zestful, on-the-go creatures seldom spend more than a couple of days in the same place. Each adult travels within its familiar home range, which may extend over 80 or 100 kilometers, using and reusing its customary lavatories, daytime dens, and the "pulling-out places" where it rolls in the grass or snow. If its route goes overland, between two ponds, for example, it will eventually trample down a pathway as it lumbers back and forth.

The long, sinuous body and stubby legs that serve the otter so well in the water are less advantageous on land. The animal's usual terrestrial gait is a heavy-footed gallop in which the back is alternately humped and extended, inchworm fashion, as the hind feet and then the forefeet are brought forward in pairs. Yet for all their apparent awkwardness, river otters can outrun people and commonly make overland excursions of several kilometers, even in the winter. By traveling on the packed snow of game trails or on ice, they are able to make the best of their low-slung forms: typically, they take several running steps to build up momentum and then plop onto their bellies for a slide, doing this over and over without loss of speed. If pursued, they can hit thirty kilometers per hour with this gait.

Like most things river otters do, this method of locomotion can be adapted for sport. Downhill sliding, as otters practice it, is a game for all ages, any number of players, and every season, since it can be enjoyed on snow-, grass-, or mud-covered banks. Down they go headfirst (sometimes on their sides or backs), to splash into the water or slither out onto the ice. Then a quick turnaround, and it's back up the hill for another run. One exceptionally enthusiastic pair was observed to make twenty-two consecutive round trips. Otters also play with toys such as twigs that can be ferried along on the end of the nose and pebbles that can be dropped underwater and retrieved.

The otters' introduction to the art of

With its water-repellent coat and sinuous form, the river otter is highly adapted for an aquatic life; yet on shore, it is active and lithe, whether traveling, playing, or simply looking about.

Rare elsewhere in western Canada, river otters are abundant along the British Columbia coast, where their principal prey is crab.

having fun comes in youthful rough and tumble with their littermates. A typical otter family numbers two or three and is born in late winter or early spring. The female usually houses her offspring in an abandoned muskrat, beaver, or woodchuck den, snuggling them into a nest of soft vegetation and wrapping herself around them in a cozy ring. As soon as the cubs have grown their waterproof adult coats, their mother assures that they learn how to swim, either by calling them to join her in deep water or by dragging them bodily in; she then acts as lifeguard for their first floundering attempts. She may also oversee their initiation to hunting, by carrying live prey into the shallows and letting it go where they can scramble after it.

A fully experienced otter is a versatile hunter. If alone, it may cruise with its eyes beneath the surface of the water, scouting for fish, and then dart straight down when its quarry comes in sight. In winter, it may dive for fish through a hole in the ice. If it has a partner, the two of them will sometimes cooperate in driving fish into shallow water and thereby each increase its chances of success. Alone or in company, otters will sometimes root in the mud on the bottom, grubbing for frogs or whatever else they can find, with their heads down and tails pointing out of the water like so many dabbling ducks. In general, they take those species of fish, amphibians, invertebrates, birds, and small mammals that can most easily be caught. In the ocean bays and harbors along the Pacific coast, for example, otters subsist on crab; elsewhere their staple is fish.

Although river otters are the most common saltwater mammal along the coast of British Columbia, including the harbors of Vancouver and Victoria, they are not abundant inland. Both their range and numbers are thought to have diminished due to human activity; people are their only significant enemies. Even in recent years, about 5,500 river otters per annum have been killed for their pelts in western Canada.

# Sea Otter

Two species of otters grace the waters along our west coast — the river otter and the sea otter, *Enhydra lutris*. Although closely related, these animals are not difficult to tell apart, even on the basis of a brief observation. Habitat alone may offer sufficient clues. Paradoxically, it is the river otter that ventures farthest out to sea: any otter that is sighted several kilometers from the nearest land, gamely swimming from island to island in a coastal archipelago, is likely a river otter. The same can be said of an otter seen fishing in fresh water or feeding on shore. Sea otters, by contrast, live exclusively in shallow, salt waters, usually on rocky, open coastline and seldom along inside passages. Although in some parts of their range they seek refuge on beaches during severe storms, they ordinarily do not go ashore.

There is good reason for this: sea otters are lubberly on land, in part because, like marine mammals in general, they tend to be heavier than their terrestrial counterparts. With females that reach more than thirty kilograms and males that weigh half again as much (or about three times as much as the largest river otter), the sea otter is the heftiest member of the mustelid family. In the unflattering but accurate estimate of one observer, a mature sea otter looks like nothing so much as a "loosely filled potato sack." The task of transporting this bulky body falls to the mittenlike forepaws and the long, webbed hind flippers, the latter, in particular, a hindrance ashore. Thus, on land, sea otters get about at a slow, hopping gallop; a slower, rolling waddle; or an even slower, belly-dragging slide.

A sea otter feels most comfortable when it

Sea otter

is lying flat on its back in the water, preferably where it can wrap up in a strand or two of kelp as an anchor against currents and tides. In this position, usually with its flippers drawn up and its forepaws across its chest, it will sleep the night away or enjoy a morning or afternoon nap. To travel, it just slips out of its weedy bedclothes and scoots off, still on its back, by paddling with its hind feet. Alternatively, but less typically, it may roll over onto its belly, once again relying on its flippers for propulsion (a sea otter does not paddle with its stubby front limbs). Underwater, it raises and lowers its flippers and tail in synchrony, using them as if they were a single swimming organ like the flukes of a whale. This is the sea otter's fastest stroke, though at a flat-out maximum of nine kilometers per hour, it doesn't set any records for underwater speed.

The sea otter's preferred foods are slow-paced creatures such as snails, mussels, sea urchins, crabs, and, sometimes, sluggish fish. These are caught on or near the bottom, usually at depths of thirty meters or less. Although sea otters are capable of staying down for five or six minutes at a time, their food-getting dives seldom last more than a minute and a half. They work quickly and nimbly, reaching into crevices, patting around boulders, feeling kelp holdfasts, or burrowing headfirst into the silt, relying on their sensitive forepaws and whiskers to locate food. As each item of prey is captured in the front paws, it is stowed in one of the otter's two capacious "cargo holds" — pouches of loose skin that extend from mid-chest to the underarms — thereby leaving the forepaws free to continue the hunt. The burden of delicacies is then freighted to the surface, where the otter can lie back, unpack, and eat in comfort, often resting its meal on its chest between bites. One otter on record was seen unloading six sea urchins and three oysters from her underarm folds.

Sea otters are also known to lug rocks up from the bottom, some of them bigger than a man's fist. Incredible though it may seem, these are placed on the chest and used as bases on which to crack large mussels and hard-shelled clams that cannot be broken with the teeth. To accomplish this, an otter grasps the mollusk between its forepaws, with the flat sides of the shell against the palms, and bashes it repeatedly on the rock. (Sometimes a second mollusk shell is used as an anvil instead.) On the average, it takes three dozen rapid, ringing blows before dinner is served. Underwater, a sea otter may use a rock to batter the base of an abalone until its quarry loosens its hold on the ocean floor. These behaviors make sea otters eligible for membership in the exclusive club of tool-using animals.

Sea otters are also remarkable for their voracious appetites. It has been estimated that a mature individual requires in excess of 3,000 calories per day, or more than enough to support a person of seventy kilograms. This is because, unlike most marine mammals, sea otters do not carry a layer of blubber to protect them from the chill of their environment. What they have instead (in addition to a high rate of metabolism) is a coat of ultra-thick, ultra-dense fur, to which they give the utmost care. While feeding, for example, they usually roll a few times to slosh off food scraps and then, if it is time for a rest, they lie on their backs and vigorously rub themselves with their forepaws, flexing their supple bodies and twisting inside their oversized hides so that no spot goes untouched. The objective of this frenzied activity is to remove water from their inner fur and replace it with air, which is essential for both insulation and buoyancy. Sometimes, as part of their preening, otters blow on their bellies or churn the water with their forepaws to aerate their coats. Then, just before dozing off, they may turn a series of shallow somersaults that leave the guard hairs smooth and filmed with water, to help prevent the underfur from getting waterlogged.

Historically, sea otter pelts fetched a high price on fur markets, particularly in the Orient. By the early years of this century, when legal protection was finally achieved, the species had been all but annihilated by hunters. Most, if not all, of the seventy-odd sea otters that now live around the Queen Charlottes and Vancouver Island are descendants of animals introduced there since 1969. Happily, recent

**With a stone resting on its belly for an anvil, this sea otter is bashing a mollusk shell to open it. The frequency of tool use by sea otters varies from region to region within their range, depending on the type of food that is locally available.**

sightings have included females with pups, so the population can be expected to undergo a gradual increase, barring some disaster, such as an oil spill. Where food is abundant female sea otters may reproduce annually, but they generally bear young only every other year. Mature animals spend most of their lives cloistered away from the opposite sex, since the males commonly congregate in certain well-defined areas, usually around rocky points, and the females disperse along the shoreline between these monkish precincts. Mating can occur in any season, whenever a male ventures into a female area (perhaps establishing a territory there) and searches for willing consorts.

The issue of such a union is usually a single pup, which is solely the female's charge. At first, the fluffy infant, unable to swim or dive, rides on its mother's chest, except when she leaves it afloat in order to dive for food. Even older youngsters that have begun to hunt for themselves evidently still require their mothers' devoted help, for when food is scarce, yearling sea otters are sometimes abandoned and these cast-offs generally die. In some places, predation by bald eagles is another peril of infancy. Although the occasional sea otter is taken by sharks or killer whales, the adults have little to fear from predators, and under ideal circumstances, they are thought to survive in the wild for up to twenty years.

**In preparation for a nap, this sea otter has anchored itself in a kelp bed. The large hind flippers serve both as paddles and, when the weather's warm, as radiators.**

# Mountain Lion

Time was when the mountain lion, *Felis concolor*, that quintessence of power and grace, ranged across North America, from the coastal mountains, through the prairies and parklands, and into the eastern woodlands. Today, following decades of remorseless pressure from bounty hunters, the species has all but vanished from most of its former haunts. In the United States, sizable populations occur only in the western mountains and in the Everglades. In Canada, significant populations survive in the mountain wilderness of British Columbia and Alberta, but elsewhere the animals are so rare as to seem fabulous. In Manitoba, for example, there are thought to be about fifty; in Saskatchewan, fewer still. Until the mid-seventies, when two cougars were shot on prairie farms, few people could bring themselves to believe that the cats were really there.

The mountain lion is superbly adaptable, demanding little more of its habitat than adequate cover and prey. Although willing to eat almost any animal food, be it a grasshopper, porcupine, or horse, its specialty is bringing down deer by stealth. The element of surprise is all important since, like most cats, the cougar lacks the high-speed endurance needed for a long chase. Instead, it generally stalks its prey to within a few meters, ever watchful against giving alarm, and then hurtles forth out of nowhere onto its quarry's back. When all goes well, the kill is clean and quick, made either by breaking the victim's neck or by biting the neck and throat; but sometimes things go wrong, as when the force of the cougar's lunge sends both hunter and hunted crashing to their deaths. Cougars have also been killed by the sharp hooves of female cattle and elk fighting to defend their calves, although such incidents are unusual. In general, if the cougar makes a successful stalk, it will make its kill, particularly since it often tips the balance in its favor by singling out a fawn or, in winter, a buck weakened by the exertions of mating time.

Cougars are potentially as dangerous to one another as they are to their prey. Imagine the results of an alley-cat-style brawl among these death-dealing cats! Not surprisingly, scraps between mountain lions are extremely rare. A female will sometimes risk combat with a tom to prevent him from killing her young (as the males are prone to do) but for the most part, cougars avoid conflict by the simple expedient of shunning society. Mature males live in virtual isolation from one another, each within an abode of perhaps 450 square kilometers. The females, which on the average confine their activities to an area only half that large, usually share parts of their living space with other adults of both sexes. Yet even if several animals occupy precisely the same range, as sometimes occurs, they almost never associate. Whenever two cats meet, one of them will generally move discreetly out of the way. At times, they avoid even this limited contact through their use of "scrapes," little mounds of soil or litter, scented with cougar urine, that are made by the males. Most of these markers are temporary, but every tom has several that are kept freshened up. Other lions go out of their way to visit these spots and will sometimes turn smartly around and head off in the opposite direction after sniffing one. Since this reaction occurs only when another cougar is already in the vicinity, it seems likely the visitor has picked up an olfactory message that translates as "Occupied. Keep out."

Mountain lion

Almost the only social encounter between adult mountain lions is mating. For the male, this happens more or less frequently, depending on the number of females that make their homes within his sphere, since each of them will usually entertain callers for about ten days every two years. Even during these brief periods, a female doesn't go overboard in putting on kittenish charms: she's apt to be snappish, snarly, and loud. Human observers describe her yowls as bloodcurdling. The male cougar, however, is not deterred, and a litter

**At birth, mountain lion kittens have mottled coats for improved camouflage. These youngsters have already started to lose their spots, which disappear entirely within a year.**

of two or three dappled, purring kittens usually appears about three months afterwards. Although litters are born throughout the year, most births are thought to occur in spring and late summer at our latitude.

From the beginning, the youngsters are solely the female's responsibility. They're a zestful lot, given to pouncing on their mother and chewing her tail. For the first few months, the female commutes between her family and her kills, hiding the former in a den or thicket and the latter under a layer of leaves or other litter. Later, as the kittens become capable of travel, they join their mother at her kills and spend increasing amounts of time prowling on their own. Then one day, usually when the youngsters are about two years old, the female walks off and leaves them to fend for themselves.

Although nearly full grown by this point, the young cats cannot yet be counted as fully mature because they don't have homes. Before cougars breed, they ordinarily must become attached to a particular neighborhood, an area in which they spend most of their time and where they know both the hazards and the best hunting sites. This is crucial for the females, which have to make regular kills while they're rearing young. But under the genetic and social conventions by which mountain lions live, an animal cannot set up housekeeping wherever it likes. Only a certain number of individuals will settle in a given region, so many males and so many females, each within its own home range. As every new generation becomes independent, its members wander widely, looking for vacancies created by deaths in the "establishment." By limiting the number of breeding animals, this system of land tenure helps keep the population in check, an important mechanism for a predator that is "top cat" on its food chain.

"Mountain devil" and "sneak cat" are two of the many colorful names that have been bestowed on the mountain lion, one of the largest and most impressive cats in the western hemisphere.

# Bobcat and Lynx

Suppose that some night, in the shadowy light of dusk or dawn, you were to catch sight of a wildcat, just as the animal turned and vanished into the bush. Depending on where you are, this stub-tailed apparition might be either a bobcat, *Felis rufus*, or a lynx, *Felis lynx*. Of the two, bobcats have the more southerly distribution. In western Canada, at the northern limit of their range, they are only occasionally seen, usually around thickets, coulees, or rocky uplands. Lynx, with their husky build, long legs, and broad, snowshoe feet — the hind paws average twenty-three centimeters across — are adapted for travel in snow. Their home is the boreal forest, anywhere from the treeline south. Thus, in thin woods and broken country, one should expect to see bobcats; in closed, climax forest, look for lynx.

There is only one exception to this rule of thumb, but it's a major one: about once every decade, lynx emigrate in great numbers from their wooded homelands toward the open spaces of the tundra and plains. In 1962-63 and again ten years later, the human-wary cats even moved into big cities like Winnipeg, Edmonton, and Calgary. Significantly, each of these mass dispersals occurred a couple of years after a peak in the numbers of snowshoe hares, the lynx's staple food. Hare populations fluctuate regularly and dramatically, with an average difference between the valleys and crests of about twenty-fold. A square kilometer

Long ear tufts, huge feet, a conspicuous ruff, and a tail with a solid black tip are among the distinguishing characteristics of the lynx, shown here. The bobcat's tail, by contrast, is marked with several black rings and a black spot near the tip.

Bobcats and lynx both specialize in preying on lagomorphs (rabbits and hares). This successful hunter is a bobcat, with the short ear tufts and barred front legs that characterize the species.

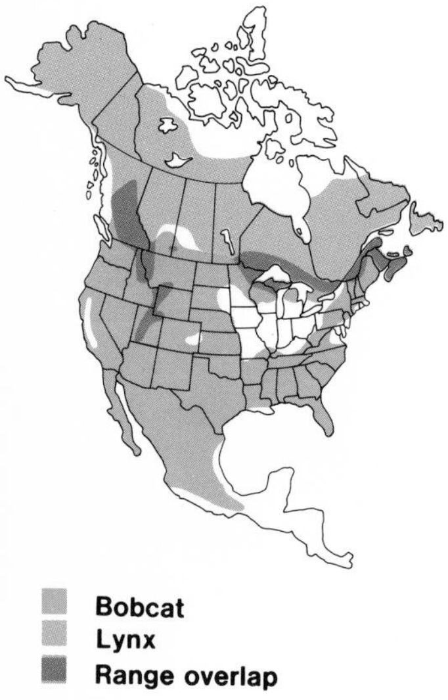


of land, which during a boom year may be hopping with hares (some 1,000 of them), will typically be down to fewer than 50 after the crash. A complete cycle takes from eight to eleven years; every decline is followed by a resurgence, every resurgence by a decline. And the hapless lynx are dragged behind, up and down this roller-coaster course. A plot of forest that supports four lynx in times of plenty may bring one of them through the lean years, or none at all. Such cyclical oscillations in lynx numbers, synchronized over vast areas of Canada, can be traced back in the fur-trade records for two centuries.

To put the situation in human terms, the lynx's economy is unstable because it is based on a single "crop." When hares are plentiful, the cats eat virtually nothing else, especially in the winter. Adult lynx kill something like 200 hares apiece every year, perhaps more when conditions are right. Whether by instinct or observation, they are experts on the habits of their prey and spend their nights prowling along hare runways or lying in ambush beside well gnawed groves, ready to pounce at the first telltale sound or sight. Bobcats, too, are skilled at hunting snowshoe hares, but their catch may include cottontails and jack rabbits as well. Lynx do not enjoy this luxury of choice. In their habitat, snowshoe hares are the only prey of suitable size to support a middle-sized cat. Lynx are generally about two or three times as big as the average housecat; hence they are seldom willing to tackle white-tailed or mule deer. Rodents and birds, though easier prey, aren't wholly satisfactory because they're too small: it takes a lot of grouse or voles to equal the weight of one hare. During hare lows, when lynx are forced to rely more heavily on such snack-sized prey, they can never get enough of them to make up for the loss.

Outright starvation of adult lynx is not thought to be a major factor in their population declines: the younger generation are the ones to suffer most. In good years, a mother lynx typically has four frisking kittens to care for, from their birth in April or May until they head out on their own early the following spring. But when food is limited, litters are smaller and fewer of the females conceive. Those kittens that are born rarely survive beyond December. This situation can persist for three or four consecutive years, with few, if any, of the youngsters reaching maturity and the older animals continuing to die, whether of natural causes or at human hands. If trappers kill too many of the adults in an area (that is, if they remove too much of the breeding stock), the local population will fall below its recovery point. In the past, heavy trapping has contributed to long-term decreases in lynx numbers and to a contraction of their range.

No one fully understands how an unmolested population of lynx compensates for fluctuations in its food supply, though it is clear that the adjustments extend to the subtleties of their social relationships. Both species of wildcat are basically solitary. Like mountain lions, they leave excremental "messages" for conspecifics, inviting them to go away. For example, they spray urine on the bushes and snowdrifts along their hunting trails, doling it out a few drops at a time; they also erect "cairns" in the form of feces displayed on stumps, anthills, and the like, to advertise their occupancy of important sites. Yet for all the energy they put into maintaining their privacy, they are able to tolerate crowded quarters when under stress. If snow conditions make good shelter hard to find, several bobcats will winter in one rock pile, each in its private apartment. They can live together for weeks without making social contact, friendly or otherwise. Lynx have never been known to do this (for one thing, they seldom use dens), but their living arrangements are nonetheless flexible. When hares are scarce, the usually standoffish cats will congregate wherever prey can still be found.

Sometimes, under very special circumstances, lynx go so far as to assist each other. One was once observed creating a disturbance in a marmot colony while its partner, unseen and unsuspected, prepared to take advantage of the uproar by catching dinner for them both. Such incidents of cooperation, though exceedingly rare, hint at a remarkable versatility of behavior among wildcats.

## Order Pinnipedia

# Sea Lions and Seals

The pinnipeds are a group of about thirty living species — all of them aquatic, most of them marine — including sea lions, seals, and the walrus. These animals can best be described as amphibious, since, though highly adapted to hunt underwater, they must periodically haul out on land (or ice), whether to rest, breed, or bear their young. This requirement for terra firma is a lingering legacy of their distant ancestors, which were land-dwelling carnivores. (Their diet is probably another inheritance, since all pinnipeds are strictly flesh-eaters.) Partly because of this descent, many zoologists prefer to include the pinnipeds in the order Carnivora, rather than assign them to an order of their own. Other biologists, pointing to the extensive anatomical changes that have occurred over the course of pinniped evolution, argue that the group is sufficiently distinct to be set apart.

Whatever the taxonomic status of pinnipeds as a whole, the membership of the three families is not in dispute. The eared seals, or Otariidae, are represented in our area by Steller and California sea lions and by the northern fur seal. (The latter species is a seasonal migrant that does not normally come ashore in British Columbia, Washington, or Oregon.) As the family name implies, this group, unlike all other pinnipeds, possesses external ears, though these have been reduced to small, head-hugging furls. This change has come about in the interests of streamlining. As part of the same general process, many of the leg bones have been contracted and withdrawn inside the trunk, with the result that little more than the feet now protrude. These appendages take the form of broad flippers, front and rear, an adaptation that is shared by all the Pinnipedia — literally, the "feather-footed ones." How the flippers are used varies from family to family. The eared seals swim by stroking with their front flippers, using their hind limbs primarily as rudders. On land, their back flippers swing forward, under the body, allowing the animals to lumber about on all fours.

The walrus, sole member of the family Odobenidae, gets about in much the same way, although it sometimes swims by stretching its hind limbs out behind, "palms" together, and paddling with lateral strokes. Walruses occur off Alaska but not British Columbia; their western Canadian range is largely restricted to the coast of Hudson Bay. Among the many peculiar characteristics of these biological oddities are their outsized canine teeth, or tusks, which are used as weapons against predators and as pitons for climbing onto ice. Hence their scientific name, *Odobenus rosmarus*, the "tooth-walking sea-horse." All other pinnipeds have unspecialized teeth that are adapted for grasping and tearing.

The remaining family of pinnipeds consists of the earless seals, or Phocidae, also known as the true seals. This group has five representatives in western Canada: three in the Arctic — ringed, harp, and bearded seals — and two in the Pacific — northern elephant and harbor seals. Harbor seals are residents at our latitude, while northern elephant seals occur only as occasional visitors. Like other true seals, these animals all swim by sculling with their hind flippers, which are permanently extended to the rear. On land, unable to use their back limbs, they may hitch along on their bellies, caterpillar-style, drag themselves forward by using their forelimbs, or combine both techniques. Despite their inability to walk, certain species of true seals can outdistance a running man.

As significant as the differences amongst the three families are, all pinnipeds exhibit many of the same adaptations for aquatic life. In addition to flippers and streamlining, these include large body mass, which makes it easier for the animals to keep themselves warm, and subcutaneous blubber, which provides padding, reserve energy, buoyancy, and insulation. To equip them for enduring long dives, they have various means of conserving oxygen, including a pulse rate that slows dramatically when the breath is held. In response to the pressures experienced at great depths, certain body structures (notably the trachea and the lungs) have evolved in such a way that they can collapse without harm.

Biologists believe that these and other adaptations have each evolved twice in the pinnipeds — once in the true seals and again, later, in the other two families. In the first case, the starting point was likely an ancient member of the weasel family, in the latter, an ancestor of the bears. Thus, the pinnipeds appear to offer a classic example of convergent evolution — animals arriving at similar solutions to common problems but by different routes.

# Sea Lions

California sea lion

The circus animals that are commonly known as "performing seals" are actually California sea lions. Free-living members of this agile species, *Zalophus californianus,* can most easily be observed on their breeding grounds off California and Mexico, but a few migratory bulls usually winter as far north as British Columbia. Here they are sometimes found in the company of the resident species, the Steller sea lion, *Eumetopias jubatus.* On Race Rocks near Victoria, for example, a California bull can occasionally be seen taking the winter air with a small herd of its northerly kin.

The visitors can be distinguished by their color, size, and voice. They tend to be darker (almost black when wet) and smaller: the males seldom exceed 300 kilograms, roughly the same as a Steller cow and about one-third the weight of a full-grown Steller bull. In keeping with their somewhat daintier proportions, California sea lions also have a more muted call — a honking bark, instead of the Stellers' deep-throated, belching roar. In addition, California sea lions are not reported to venture north of Vancouver Island; Steller sea lions, on the other hand, winter throughout the inlets and inside passageways along the British Columbia coast and north to the arctic ice.

There are many gaps in our knowledge of Steller sea lions. Little is known of their fall and wintertime activities, although, when the weather is fine, the animals can often be seen loafing on the rocks. Details are also lacking about their life in the ocean at any time of the year. Generally, they are sighted within twenty-five kilometers of land, traveling either alone, in small parties, or in rafts of several hundred or more. Group members often dive and surface in unison, presumably so that a sudden movement by one animal won't ruin the fishing for them all.

Another point on which our information is incomplete is the sea lions' impact on human fisheries, particularly the salmon catch. Sea lions are attracted to fishing boats and have a bad reputation for raiding nets and hooks. What's more, they are often seen at the mouths of rivers during salmon runs. Over the years, thousands of Steller sea lions have been killed off British Columbia to protect the fisheries, a task made simple by their sociable habits. In the spring, the "bachelors" (males up

At home in two worlds, Steller sea lions generally frequent inshore waters and bare, sea-washed rocks. Like other pinnipeds, they are more gregarious than most terrestrial mammals.

**Steller sea lion**

to five or six years of age), together with a few barren cows and mature bulls, congregate by the dozen in "non-pupping" colonies. The rest of the population gathers year after year on certain remote, sea-lashed rocks to bear their pups and breed. In British Columbia, there are two major "rookeries," one at the northwest tip of Vancouver Island and the other off the southern end of the Queen Charlotte chain. Although quick to seek refuge in the water at other times of the year, mature sea lions at breeding colonies often stand their ground, roaring out their protests and making easy targets of themselves. Thus, in the late 1950s, fisheries personnel were able to kill half the British Columbia sea lions in less than five years. Unfortunately, this "shoot first, ask questions later" approach proved ineffective. Subsequent research suggests that sea lions generally eat very little salmon and have no significant effect on any commercial fish stock. Their staple foods are "scrap" fish, herring, octopus, and squid.

For some reason or other, Steller sea lions also frequently ingest stones, some of them up to twelve centimeters across. An animal may carry as many as ten rocks in its stomach at one time. Some observers suggest that sea lions need the weight as ballast when they dive. A more plausible explanation is that the rocks serve as food-grinders. Sea-lion teeth are poorly adapted for chewing; thus fish are either ripped and shaken into bits or swallowed whole. In either case, the food may have to be pulverized before digestion can proceed. How better to accomplish this than by having millstones pounding around inside the stomach?

Yet another theory contends that the rocks distend the paunch and prevent the animal from feeling hungry during a prolonged fast, which sea lions (like other pinnipeds) frequently endure. A Steller cow, for example, typically stays with her newborn pup for two weeks before resuming her nightly feeding trips. And at breeding time, a mature bull sometimes goes two months without food, obtaining calories and fluid from his blubber.

This sacrifice allows the bull to maintain a round-the-clock watch over his breeding territory. Every year in early May, the males heave themselves out on the rookery rocks and partition them into exclusive domains, a process that involves much bluff and bellowing and some loss of blood. Territories vary in area, but, according to one study, the average runs to 250 square meters. A major preoccupation of a territorial bull is defending the cracks, ridges, and imaginary lines that constitute his boundaries, against incursions by neighboring males and by ambitious bulls that have been unable to secure territories. Sometimes a bull claims the very same area in successive years.

Why this tenacious defense of a cramped and unprepossessing piece of real estate? The answer is that along with possession of a territory comes the chance to reproduce. Pregnant cows begin to haul out at the rookeries at the end of May and congregate around the territorial bulls in so-called "harems." This is a misnomer since the ponderous males are generally not able to constrain the cows. At the height of the breeding season, in July, there may be as many as ten cows for every dominant bull, but the number varies from territory to territory and from day to day, as the females come and go across boundary lines. A cow may bear her pup inside one territory, copulate with the overseer of another, and rear her youngster within others still.

One benefit of the territorial system (apart from ensuring that the fittest bulls sire the most pups) is that it helps to maintain order in the hurlyburly of the rookery. Unrestrained conflict among sea lions is rare. Neighboring males usually settle their disputes through ritualized displays, in which the two mammoth contestants hurl down on their bellies on either side of the disputed line and menace each other with gestures and sounds. This suppression of violence decreases the likelihood of serious injury to the bulls and reduces the chance that a pup will be squashed by a lumbering male, though such mishaps do occur. Other accidents, storms, desertion, and disease also take a toll on the pups. Apart from killer whales, large sharks, and people, the adults have few enemies and may live to breed beyond their twentieth year.

The massive forequarters and bulbous necks of the two sea lions in the center and the one sitting up at the right suggest that they are bulls. At maturity, male Steller sea lions may weigh more than the largest bears.

# Harbor Seal

Harbor seal

The harbor seal, *Phoca vitulina,* has an enviable reputation for tranquility. A typical, sleepy-headed individual spends about half of its life resting on land, usually in the company of several dozen other lolling lie-abouts. Unlike sea lions and sea otters, harbor seals apparently cannot sleep while afloat, although they do nap on the bottom in shallow water, surfacing to breathe at intervals of five minutes or more. Ordinarily, however, they rest on tidal flats or exposed reefs (seldom more than fifteen kilometers from shore), taking the double precaution of selecting places that are safe from sneak attacks by land predators and within easy access of water that is deep enough for a quick escape. Inland, where seals have moved up coastal rivers to follow runs of salmon or eulachon, they haul out on sandbars, islands, and isolated shorelines in order to doze or to suckle their young. (Harbor seals are seasonal visitors to many inland waterways and permanent residents in places such as Harrison Lake, British Columbia — 180 kilometers upriver from the coast.)

For all their apparent languor, harbor seals are actually wary animals, as anyone who has tried to get near them in the wild will testify. Perhaps they inherited this trait from their distant ancestors, which are thought to have been otter-like creatures that lived in dread of terrestrial carnivores. Another possible legacy of this fear-filled past can be seen in the haste with which a cow seal gives birth to her pup. The delivery takes place on land (sometime between May and September), usually after a labor of thirty minutes or less. Out slides the pup, to lie scrawny and pitiful on the chill mud of the hauling ground. Mother and young sprawl as if stunned; then, after a few moments of recovery, they rouse and touch noses, beginning to learn the odors by which they will recognize each other, first in the "nursery" where cows with very young pups congregate to sleep and, later, on the general hauling grounds. A few more minutes pass quietly, as the infant attempts to nurse or the cow nuzzles and caresses her new offspring. The mother makes no attempt to lick her newborn or to clean up the debris of birth, as most other mammals do. Instead, she and her pup head out to sea, often within an hour of the first labor pang.

Seal pups can swim and dive at birth and are capable of surviving in the water for hours at a time. This is essential because most harbor seals live in inshore areas where their hauling grounds are flooded twice daily by rising tides. Thus, the seals' schedule of sleep and activity is imposed on them by the sea. When there is nowhere else to rest, a tired pup will climb onto its mother's back, using her shoulders as a traveling hauling ground. Sometimes, a newborn gets its first experience of diving while riding piggyback. After every dive, a mother seal sniffs her pup's nose to make sure that the youngster that came up with her is the same one that went down. Ever attentive, the cow confirms her pup's identity dozens of times each day.

In other ways as well, a seal cow proves herself to be a devoted guardian. If a pup goes astray, its cry of distress brings mother hurrying to the rescue, even when that means risking her own life. Should a curious young seal venture within gunshot of a boat, for example, the gallant cow dashes in, grabs the bleating adventurer in her mouth, and drags it, gargling, under the waves. So intent is she on keeping her youngster at her side that she scarcely eats during its first month or six weeks of life. As she gets thinner and thinner, the pup gets round and fat, nourished on her milk, which is considerably richer than whipping cream.

The cow abandons her weanling pup when it reaches a month or two of age and is soon ready to breed again. Courting couples can sometimes be seen in summer and fall as they frolic together in the sea, rolling, nuzzling, splashing, even leaping into the air. The female also vents her passion with savage bites to the

**With a companion in the background masquerading as a rock, a harbor seal cow and pup enjoy a final few minutes of relaxation before the incoming tide inundates their hauling ground. The adult's flippers are clasped and raised in a typical resting posture.**

bull's shoulders and neck, leaving him badly scarred. Little else is known about the seals' mating behavior because their streamlined, unisex contours make the partners hard for an observer to distinguish at any distance.

Although no one can say for sure, the liaison between a breeding pair is probably brief, and other associations among adult seals also appear to be casual. As far as anyone has been able to ascertain, the herds that huddle together on the hauling grounds are generally unstructured and impermanent. For the most part, a harbor seal seems willing to bed down with whatever individuals happen to be in the neighborhood at nap time. All but oblivious to its bedfellows, it rarely honors them with so much as a grunt or a surly wave from a front flipper. The rest period at an end, the animals disperse into the ocean to feed in solitude.

Harbor seals subsist mainly on fish, with an occasional mollusk or crustacean for variety. As a rule, they favor relatively slow-moving species that occur in coastal waters — herring, tomcod, flounder, and rockfishes. Given the opportunity, seals will feed on salmon as well, either intercepting them as they swim upriver to spawn or filching them from gill nets. Because of this competition with human fisheries, harbor seals were put under a bounty and persecuted for decades with hooks, traps, nets, harpoons, guns, and dynamite. Every year for half a century, from 1914 to 1964, an average of 3,000 seals were slaughtered off the British Columbia coast. Although limited hunting still occurs, the seals' major predators now are thought to be sharks and, more importantly, killer whales. Unfortunately, mankind may shortly regain precedence as a cause of death, since some scientists suspect that polychlorinated biphenyls (PCBs) and other pollutants are responsible for the high rate of birth defects and miscarriages that is currently observed among certain populations of Pacific harbor seals.

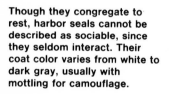

Though they congregate to rest, harbor seals cannot be described as sociable, since they seldom interact. Their coat color varies from white to dark gray, usually with mottling for camouflage.

## Order Artiodactyla

# Cloven-Hoofed Mammals

The artiodactyls, or cloven-hoofed mammals, are a comparatively large and diverse order. There are about 150 species worldwide including, among others, the camel, the hippopotamus, and the giraffe. In western Canada, we have several domesticated species (pigs, cows, goats, and sheep) as well as eleven wild ones — wapiti, mule deer, white-tailed deer, moose, and caribou, which are known collectively as cervids, or deer; pronghorns, which form a family all by themselves; and bison, mountain goats, muskoxen, bighorn sheep, and thinhorn sheep, which make up the bovid family.

These three families can most easily be distinguished on the basis of their antlers or horns. Amongst the deer, for example, it is usual for males to have antlers of solid bone that they grow anew each spring and summer and shed every winter after the mating season. Amongst the bovids, on the other hand, both sexes commonly have horns of bone covered with a sheath of keratin, the same super-tough substance from which our fingernails are formed. In adults, neither the bone core nor the sheath is ever shed. But in the pronghorns, which also have horns, the bony core is retained for life, while the keratinous sheath is discarded and replaced every year.

Whatever the differences in their headgear, there is a basic similarity in the structure of their feet. These are the *cloven-hoofed* mammals, after all. Whether you study a giraffe, a hippo, a pig, or a moose, you will find that the first digit (the one corresponding to our thumb or big toe) is missing, while toes two and five will typically have been reduced in size to form the dewclaws. The animal's weight is borne on its third and fourth digits, which have elongated to form a two-lobed hoof. (There is another order of hoofed mammals, typified by the horse, in which the third digit has evolved into an undivided hoof. These are the perissodactyls, or "odd-toed ungulates.") Cloven hooves are an important adaptation because they make an animal fleet and sure of foot and thereby permit mature, healthy individuals to escape from predators.

The artiodactyls have also evolved specialized digestive tracts. This has been necessary because they are browsing and grazing animals that subsist almost entirely on plants. Cellulose, which is a major constituent of plant cell walls, is readily digested by certain microorganisms but cannot easily be broken down by an ordinary mammalian stomach. One group of cloven-hoofed mammals (a group that includes all the wild artiodactyls in western Canada and adjacent areas) has solved the problem by developing an extra-large, four-chambered stomach, with a captive population of protozoans and bacteria in the first pouch, or rumen. These animals — the ruminants — swallow their food half-chewed and let the microorganisms work on it awhile before regurgitating it as a cud, chewing it thoroughly, and passing it back through the rest of the digestive tract. If the microorganisms in the rumen die for some reason (from a lack of protein, for example), the host animal will starve to death, even though its paunch may be full of undigested food.

This symbiosis between microorganism and mammal permits the ruminant to obtain more nutrients from each hectare of habitat than other ungulates can, with the result that a given piece of land will support more cud-chewers than other hoofed animals.

Ruminants are the most advanced of the cloven-hoofed mammals, the culmination of a line of evolution that began more than 38,000,000 years ago. At one time or another in their history, the artiodactyls have been the dominant herbivores on every continent they inhabited. In western North America, not so very long ago, they numbered into the hundreds of millions, and even today, though under continuing pressure from man, at least one species survives in every major ecological zone from the short-grass prairies up to the mountain crags and north to the arctic coast.

# Wapiti

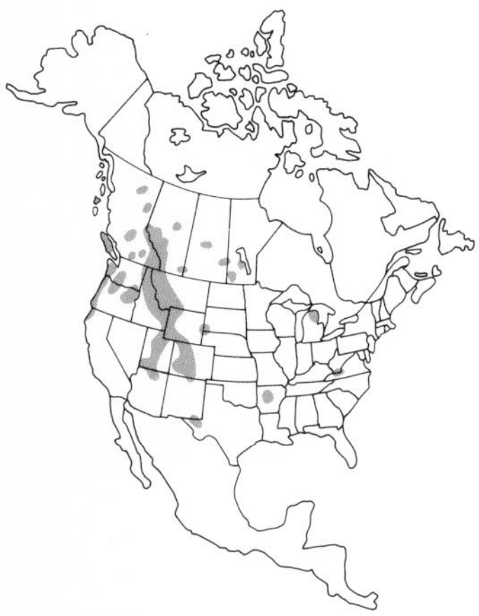

Wapiti

When the Europeans first arrived in North America, the wapiti, *Cervus elaphus,* was among the most widely distributed of the cloven-hoofed mammals. A highly adaptable species, able to utilize open or closed habitat and to consume both forage and browse, elk could then be found on mountain meadows, in mixed-wood forests, and sometimes on the plains, between the Alleghenies and the Pacific coast. During the wildlife slaughter of the last century, however, the eastern herds were exterminated in Canada and the United States.

In western Canada, the species fared only slightly better. Of the populations that once ranged from southern Manitoba to the parklands of Saskatchewan and Alberta, only a remnant remains, now confined to a few parks and wilderness areas. The Roosevelt elk — formerly resident in the lower Fraser Valley — survives only on Vancouver Island and the Olympic Peninsula. And the familiar Rocky Mountain wapiti, which has been eliminated throughout the northern reaches of its Canadian range, almost met the same fate farther south as well: most of the elk in Jasper and Banff Parks, for example, are descendants of animals imported from Wyoming after the Canadian herds had almost been wiped out.

It was the gregarious habits of the wapiti that made them such easy targets for nineteenth-century hunters. For much of the year, the females and immature animals of both sexes congregate in bands of from half a dozen to 1,500 individuals, with the largest aggregations being formed in the wintertime. In some populations, these cow herds appear to be in constant flux, with the size and composition of groups changing from day to day, even from hour to hour. In others, the herds are stable and cohesive, so much so that a cow may spend most of her life in the same, small, closely knit band (perhaps forty-five animals or fewer) to which her mother belonged.

Within these relatively permanent groupings, interactions amongst individuals are regulated by a "peck order" of dominance relationships: the older and larger an animal, the higher its rank is likely to be. A cow will give up a favored resting place when threatened by one of her "social superiors" but will still feel free to boss her subordinates. This dominance hierarchy is probably determined in infancy; a youngster's position may depend both on its mother's rank and on its own success in the chasing and challenging games of the calves. By six months of age, a young wapiti knows its place in the herd.

Under certain circumstances, a mature female will act as leader of a band of cows and calves. In the mountains, for example, the leading animal may set the pace on the spring and fall migrations, as the herd treks from winter to summer range and back again. When the animals come to a river crossing, the leading cow may decide that she's in no hurry for a swim, and all the others will await her pleasure. A few high-spirited yearlings may head out on their own, but they are sure to lose their nerve and turn back. This can go on for hours or even days, without any need for

Majestic though his antlers are, the wapiti in the foregound has only five points on each side, giving him the typical rack of a three-year-old. (Older bulls usually have six points.) The animal in the background is shedding his velvet.

**Dappled and leggy, wapiti calves are born in late May or early June and spend most of their time in hiding for the first week or two. Single births are common among wapiti, although twins also occur, especially on good range.**

violence to maintain discipline.

The order of the cow herds is temporarily disrupted each fall by the rut. For much of the year, the mature bulls (those over three years of age) live in small, all-male groups. But come September, they begin to feel hostile toward their "buddies" and eager for female company. Unlike most other male mammals, which must locate their mates one by one, a bull elk can hope to consort with a whole herd of females at the same time. In theory, this gives him the perfect opportunity to mate with each doe as she becomes fertile. In fact, life is seldom so simple as that. First, the bull attempts to drive the immature males (all but the current year's calves) out of the cow herd. Then he must contend with any other mature bulls that happen to be hanging around — and there may be several. Back and forth the herd bull must go, sparring here, chasing there, threatening runaway cows. Sooner or later he will inevitably lose some or all of the females to one of his challengers. During the rutting season, the cow herds are usually subdivided into groups of from five to fifteen females.

Obviously, the bull that is best able to intimidate or overpower his rivals will have access to the most females for the longest period of time, since he can keep other would-be "harem masters" away. This must make life frustrating for the losers, but it benefits the species by ensuring that only the most vigorous males get a chance to mate. However, if such conflicts routinely ended in the death of one or both contestants, the evolutionary advantage would be muted, to say the least. This is why bull elk, and other male cervids, have acquired complexly branched antlers. A jousting bull's first impulse is to gore his opponent on the flank. Rather than let that happen, the threatened animal will meet the attack with his own well-armored head. Then, their antlers securely interlocked, the contestants can heave at each other eyeball-to-eyeball. In this way, they are able to indulge in a wholehearted shoving match — sometimes lunging with enough force to snap off bits of antler — and yet generally avoid fatal injury, since neither animal is free to deliver a side-raking slash.

Most observers agree that, among wapiti, really determined battles over access to females are uncommon. Usually the question can be settled with a few intimidating gestures and calls. But in spite of this restraint and the safeguards provided by the antlers, casualties do sometimes occur. After all, a large bull wapiti, at a weight of half a tonne, is a fearsome adversary. His antlers alone can account for ten kilograms or more.

This regal headdress serves the bull well during the mating season, not only in contests with other males, but in such seemingly mundane pursuits as thrashing saplings and digging up the ground. For a long time, it was assumed that wapiti and other deer sparred with bushes in order to rid their antlers of "velvet," the vascularized skin that nourishes the bone while it grows. In fact, the main function of this behavior is to rub bark off the trees, thereby creating a marker that is detectable by sight and, perhaps, by smell. The scrapes on the ground are thought to function in much the same way, for the bull not only paws and plows the ground, he also urinates on the freshly worked dirt. (He may also make a visual and olfactory "marker" of himself by rolling in a wallow of urine-laced mud.) Whatever medium he chooses, the message is always the same: "Attention all male wapiti: sexually aroused bull nearby." It is an invitation to a tussle over dominance.

Wapiti also communicate with each other through their "roaring," or "bugling." Although there is much variation in these calls, they typically start as a low-pitched moan, swoop up into a clear, ringing cry, and then subside with one or more cough-like grunts. A bugling animal stands with its muzzle raised, rather like an impassioned opera singer at center stage. For some unknown reason, female wapiti occasionally bugle in the early spring, around the time they give birth to their calves. The males' full-chested bellows are most often heard in the fall; apparently they help bulls find one another for tests of strength. A population of wapiti bulls in full chorus can be pandemonium. One biologist, who spent what must have been a hair-raising night in the forest, counted up to forty "roars" within a five-minute period. Small wonder that wapiti are said to be the noisiest of our deer.

# Mule and White-Tailed Deer

Technically, caribou, wapiti, and moose are all "deer," but in everyday speech, the term is reserved for two closely related species — the white-tailed and mule deer, *Odocoileus virginianus* and *Odocoileus hemionus*. At first glance, these animals may seem bafflingly similar, particularly since that first glance is usually all one gets. Deer, especially the high-strung white-tails, are wary creatures: their best defense against predation lies in avoiding detection, and failing that they flee. Except in the spring, when the demands of pregnancy force the does to eat during the day, all deer do much of their feeding in the half-light of dawn and dusk, when they are difficult to see. So it is no wonder that many people confuse white-tails and "muleys".

A century or so ago, the question would seldom have arisen. Until the 1880s, the prairie provinces were largely the domain of the mule deer. Then, as human settlement proceeded west, the white-tails did too, to take advantage

of the sheltering bluffs that grew up as prairie fires were brought under control. At present, both species appear to be expanding northward, particularly the hardy mule deer. Just why this should be is not clear, though it may be an unintentional result of human activity. Lumbering and fire both create habitat for deer, by opening the forest canopy and letting sunlight through to the ground. This promotes the growth of short plants and brush on which deer browse. Whatever the reason for the shift in populations, both species may now be found in many regions of western North America. White-tails are more common in settled areas, while mule deer prefer forest margins and broken countryside. One of the muleys' adaptations to life in rough terrain is the characteristic bounding gait that they use when alarmed, an attribute that has earned them the nickname of "jumping deer."

Once you know what to look for, it is fairly easy to distinguish between the two species on the basis of appearance. If the first thing you notice is a pair of broad, mule-like ears, then you've come across a mule deer. But if the animal takes fright and sprints away, flashing the underside of a flaglike tail, some thirty centimeters in length, you have found a white-tail. (The mule deer's tail is shorter and marked with black; some subspecies have conspicuous white patches on their rumps.) During much of the year — from summer through midwinter — large bucks can also be identified by the form of their antlers. On a typical white-tail, the tines, or "points," simply extend upward from the main beams; on a mule deer, they extend upward and then fork.

In either species, antlers can reach cumbersome proportions. Six to eight points is considered a good-sized rack for a white-tail, but the record-holding animal boasted a total of fourteen, displayed along main beams that were each over seventy centimeters long. A buck that produces a large rack may be under as much strain as a doe that has twin fawns, for if his diet is inadequate, calcium will be withdrawn from his skeleton. Yet antlers are not useful as a defense against predators (both male and female deer lash out with their front hooves if sorely pressed), and they may constitute a direct hazard to their bearer: occasionally they get tangled in trees or in the antlers of

Facing page: This white-tailed buck is wearing his reddish summer coat, which, later in the season, will moult to grizzled gray. The antlers, still in velvet and not fully grown, already display the characteristic white-tail form, with tines that extend up from the main beams but do not fork.

The high-flying gait of a frightened mule deer permits it to clear obstacles and follow an unpredictable, zigzag course in evading predators.

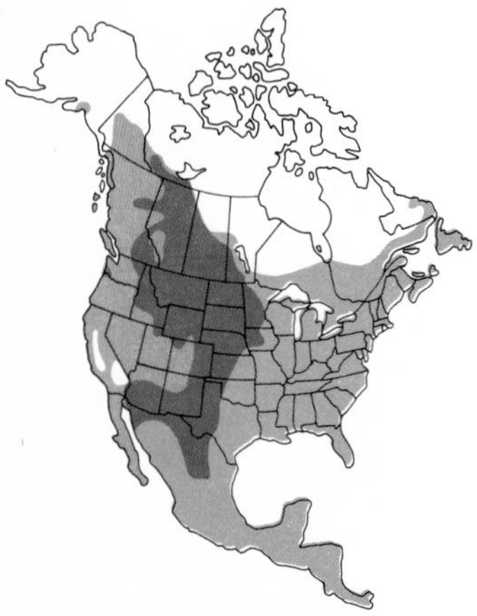

Mule deer
White-tailed deer
Range overlap

another animal. This means certain death, either through predation or starvation.

If individual survival were all that mattered, it wouldn't make sense for a buck to sprout all this bone every summer, lug it around through half the winter, and then drop it in a snowbank. But antlers prove their worth in late fall and early winter, during the mating season, when the bucks test their reproductive "fitness" in head-on combat. By far the majority of interactions are simply sparring matches — tussles between bucks of different size that have previously determined their relative ranks. Under these circumstances, the outcome is never seriously in doubt, and the bouts take on the half-playful atmosphere of competitive sport.

All-out battles between evenly matched males, in which dominance is at stake, are both dangerous and rare. In many instances, rival bucks can settle their differences without resort to force. A belligerent white-tail may first signal his aggressive intentions by laying his ears back and then lowering his head as he gives his opponent a hard stare. If that doesn't work, he will sidle toward his adversary with head high and hair raised, making himself look as ominous as possible. Finally, as the ultimate threat, he presents his rack. A deer that witnesses this performance and decides he's been outclassed will quickly leave, and a wise person will immediately do likewise, for the next step in the sequence is a sudden, headlong charge, in which there is real danger that the opponent will be stabbed.

In these confrontations, deer evidently communicate with one another by means of visual cues, but they are also capable of interacting through their acute sense of smell. Both species have "tarsal glands," conspicuous tufts of hair on the insides of their hocks; these patches are larger in dominant animals than in subordinate ones. The primary function of these patches is to hold the scent of glandular secretions and of the urine that each individual carefully deposits on its hocks. These odors are apparently involved in the animals' dominance relationships, for when two deer meet, the subordinate animal sniffs the hocks of its social superior. Deer also have "inter-digital glands" between the lobes of their hooves that deposit scent with every step, leaving trails that help the animals track each other's movements.

The females, for example, use these scented footprints to locate wandering fawns. A mature doe will typically produce two dappled offspring in late May or early June. No sooner have they emerged from the womb than the mother sets briskly to work, licking them clean and odor-free. She then leads them away from their birth place, where telltale fluids may have soaked into the ground, and caches them, each in its own dusky hiding place. People sometimes make the mistake of "rescuing" these gangly youngsters, assuming them to be orphans. In fact, the mother has probably moved off to browse until time for her young ones to nurse. Should a fawn ramble away on its own while the doe is gone, she will track it down by following its unique scent trail.

The mating season is another time when deer locate each other by smell. A rutting white-tailed buck will make a series of "scrapes," small patches of ground, pawed clear of litter, on which he urinates. Now and then, he will return to check these spots for the scent of doe urine (passing females also urinate on the scrapes). If the smell is promising, the buck sets his nose to the ground and snuffles off eagerly along the doe's scent trail. Biologists haven't decided if this is the only function of scrapes, but it does appear that they function as Lonely Hearts Clubs for white-tailed deer.

Chances are, the buck will find the doe nearby, for deer are not great travelers as a rule. About the only time when they undertake long journeys is if they are forced into them by bad weather. In the wintertime, white-tails may travel several miles to a gully where they can congregate, or "yard up," out of the chilling winds. In the mountains, herds of mule deer migrate up and down the slopes as the season dictates, descending from the alpine meadows in the spring and returning in the fall. They follow the routes chosen by their ancestors, each generation showing the way to the next.

On a typical mule deer buck, such as the one shown here, the tines of the antlers are forked. The relatively short tail (twenty centimeters or less), marked with black, is another distinguishing characteristic of the species.

# Moose

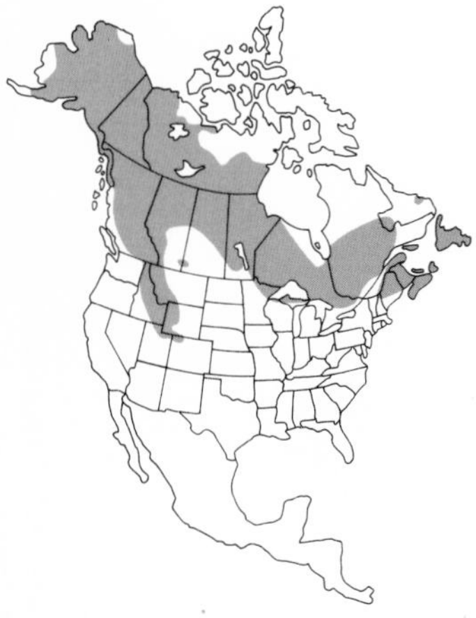

Moose

The North American moose, *Alces alces,* is the largest deer in the world. The bulls average about half a tonne (as much as the biggest wapiti) and sometimes reach 850 kilograms. Though the cows are generally more lightly built than the males, a large adult of either sex stands almost two meters high at the shoulder.

For an animal that is so imposing at maturity, the moose is surprisingly tiny at birth. Newborn calves range from about fifteen kilograms, for a single birth, down to a mere six or seven kilograms each, for some sets of twins. But moose don't stay in the lightweight class very long: for the first several months of life, they grow at a frantic and accelerating rate, gaining as much as two kilograms a day. No other cloven-hoofed mammal can put on weight so fast.

Given this rate of growth, it shouldn't be surprising that young moose are very enthusiastic about food. A nursing calf isn't content to stand quietly and suckle; instead it bunts eagerly at its mother's udder all the while it sucks. With a newborn, this is fairly gentle — just a rhythmic succession of taps to speed the flow of milk. But as the calf grows stronger, it becomes more insistent, until by the time it's two or three weeks old, it is delivering a barrage of solid blows. Each thrust — and they come once or twice a second — is hard enough to set the calf's ears flapping wildly. When the cow decides she's had enough, she may terminate the feeding by walking away. The calf usually has such a determined grip on the nipple that it gets dragged along for a moment as she leaves.

When the cow thinks it's time for her calf to nurse again, she can issue the invitation by more subtle means. Moose communicate, in part, through body language. If the cow stands with her head high and her ears pointing out, the calf knows that it's mealtime. But if the mother is alert because she suspects some danger is nearby, she will hold her head in the same position and stick her ears up instead of out. The calf, not being too discriminating and generally hoping for the best, will sometimes come running to be fed in response to such a warning. It may not get what it wants, but at least it's where its mother can protect it if need be.

The bond between a cow moose and her young is extremely strong. Unlike a white-tailed doe, for example, which hides her fawns and then takes off for hours, a moose rarely leaves her offspring alone. The calf, for its part, has a strong, inborn tendency to heel and will trail its mother almost anywhere. Unfortunately, a very young calf is also prone to following other moose, and even people, if it gets the chance. This may explain why the cow has a strict policy of repelling all intruders, including her own kind. In general, moose are the most solitary deer in North America.

When a moose decides that someone or something has come too close to her youngster, she adopts a threatening posture — head low, ears down, nostrils flared, hair raised. This is the moose equivalent of shaking a clenched fist. If the interloper doesn't leave, the cow strikes out with her lethal front hooves. She can also connect with her hind feet when she chooses to. Thus armed, a cow can repel not only humans and moose, but wolves and other predators as well. There is a documented case of a moose that literally rescued her calf from the jaws of a large black bear.

The only time of year when a cow with a calf will tolerate company is during the mating season, which generally occurs in the fall. Then she actually solicits the attention of the bulls. Her plaintive quavering call is an open invitation to any male within the three- to four-kilometer radius in which it can be heard. But if she's successful in attracting a mate, she will stay with him only if he shows no aggression towards her calf.

When all goes well, the bull remains with the cow throughout her fertile period, follow-

Single births are the general rule among moose, but twins are not uncommon. This mother will tend her offspring with great solicitude during their first year of life.

ing her like a shadow until she is ready to stand for mounting. If they encounter another bull in the meantime, the two males may lock antlers for a back-and-forth pushing contest while the cow and calf browse nonchalantly nearby. But if the mating pair encounters another female, then the cow will sometimes go into action. At first, such a confrontation between two females follows the usual course of intimidation and attack, but if the animals enter a freshly made moose wallow, their tactics change. (A wallow is a urine-soaked patch of mud that the bulls make during the rutting season and in which both sexes roll as part of their mating behavior.) In a fight between cows, one of the combatants will stretch herself out flat in the mud, determined to stay in charge of it. The other cow, meanwhile, will do everything she can to drive the wallower out. And through all this the male stands close by, holding himself broadside to the cows, as if to show off his physique.

Even when a female is doing battle, her young one does not leave her side. When she bristles, the calf bristles. When she wallows, the calf plasters itself to her flank. In fact, the cow and her offspring usually stay together not only through the rut but all winter long as well, for that is the season when the mother's protection becomes crucial to the calf's survival. Under most snow conditions, an adult moose is able to move quite freely, thanks to its stilt-like legs, but the calf, still not full-grown, is severely handicapped. Since the calf cannot

**Moose calves develop rapidly. Though helpless at birth, they can outrun a person and swim when only a few days old.**

flee through the drifts, it needs the cow's loyal defense.

The break between mother and young comes with wrenching suddenness in the spring, when the cow drives her yearling away to make room for the new season's calf. The yearling retreats and returns, retreats and returns, only to be forced off each time. Eventually, it gives up and wanders into the forest, perhaps to form a temporary alliance with a bull or with other displaced yearlings. In time, it will establish a home range of its own, perhaps even two — one for summer and one for winter.

This dispersal of the yearlings may be the species' way of coping with habitat that comes and goes. Though moose live in the boreal belt, they do not thrive in closed coniferous forests; rather, they seek out clearings where there is enough sunlight for the aspens, birch, and willows on which they prefer to browse. The difficulty is that these clearings are only temporary, for the conifers always creep in from the edges and take over. In the past, new openings were continually being created by forest fires. Since the activity of fire is unpredictable, the challenge for the moose was to locate this newly created habitat, and the wanderings of the yearlings accomplished exactly that. Nowadays, with fire more or less under human control, moose habitat is no longer produced at the natural rate. Strange as it may seem, Smokey the Bear is no friend of the moose.

**By late August or early September, the antlers of bull moose have attained their full size. The covering of velvet (through which the bone had been nourished during growth) dries and is shed at this time.**

# Caribou

Caribou

In fresh autumn pelage, this barren-ground caribou buck displays a handsome rack, still in velvet. Most female caribou also have antlers, but theirs are simple and small, similar to those borne by one- and two-year-old males.

Every year, with the first signs of spring, the barren-ground caribou, *Rangifer tarandus groenlandicus,* begin to congregate and move north, out of the spruce forests in which they spend the winter. Gregarious by nature, they mass first in small groups, then in herds of several thousand, and finally in companies of 50,000 to 100,000, each extending for perhaps 500 kilometers along the advancing front. The pregnant does lead the way, with the barren does and bucks straggling behind them for 300 kilometers or more.

From mid-April to mid-June, these vast herds surge toward their summer range on the tundra. Seldom deterred by weather or physical obstacles, the animals average about fifty kilometers a day, following water courses and eskers; navigating featureless, frozen lakes; and swimming ice-free streams. Usually they choose a route that is deeply scored by generations of caribou hooves, but sometimes, inexplicably, they will strike out on a new course. By some unknown mechanism, they always manage to home in on one or another of the inhospitable arctic uplands where their calves are born.

After a brief summer on the tundra, where they are often tormented by biting flies, the animals reassemble for the mating season and the unhurried fall migration back across the timberline. In the boreal forest, where the conifers break the wind and keep the snow from crusting, they pass the coldest months of the year in small, loosely knit bands, foraging for lichens, sedges, Labrador tea, and other vegetation. But as the days lengthen in March, the caribou begin to act restless again, and soon it is time for them to flood out across the tundra toward their traditional calving grounds once more.

Even today, this annual cycle of migrations is spectacular, though the number of animals involved is probably much lower than it used to be. As late as 1900, there are thought to have been 2,000,000 to 3,000,000 barren-ground caribou in Canada. By 1950, the population was down to around 670,000, and in the next fifteen years, it continued to drop calamitously, bottoming out at about 200,000 in 1964. Although it has since rebounded somewhat in response to management, there is still ample reason for concern, particularly in view of the escalating pace of development in the North.

No one knows for certain why the barren-ground herds declined, but most researchers currently think that people were largely to blame. Two factors that may have been involved are overhunting, which was made possible in recent times by the introduction of firearms, and forest fires set by human carelessness. Although fire is a natural feature of the caribou's environment and one with which they have coped for millenia, any marked increase in the area burned, over what naturally occurs, can be deleterious. A single, runaway blaze can ruin vast areas of the caribou's winter range for a hundred years or more, because lichens, the animals' staple food, are slow to regenerate. Caribou eat so many lichens — up to about five kilograms per animal per day — that their favorite ground-hugging varieties are often referred to as "reindeer moss." (Eurasian reindeer and North American caribou belong to the same species.) In spite of their hearty appetites, caribou rarely overgraze their range under natural conditions. They are nomadic creatures, constantly on the move, not only from day to day but from minute to minute. They habitually "eat on the run," selecting a bite here, a bite there, and seldom grazing over an area twice in the same season. With a mainland range of some 2,000,000 square kilometers, they have plenty of pasture from which to choose. Counting their migrations, barren-ground caribou routinely cover hundreds — even thousands — of kilometers every year.

The barren-ground herds constitute one of four Canadian subspecies of caribou. Grant's caribou, which live in the northwestern arctic,

**Caribou outclass all other members of the deer family in swimming ability. They can cross a kilometer of open water in less than nine minutes and have been known to swim six kilometers against strong currents. These bucks are on their fall migration.**

also make seasonal treks from forest to tundra, but the Peary caribou of the arctic islands remain on the barrens year round. The woodland caribou of the boreal forest and mountain regions may migrate only a few miles, up and down a mountainside, for example, as the climate demands. Still, even the most sedentary of caribou is far from settled: it is in their nature to stay on the move.

There are few types of countryside that caribou cannot cross. They are strong swimmers, thanks to a dense coat of exceptionally buoyant hair and to their scoop-shaped hooves. (These same hooves make excellent snow shovels in the winter, when the animals have to dig for food.) They negotiate soft snow or muskeg by splaying their hooves and settling onto their dewclaws, which are unusually long and wide-set. This spreads their weight over a larger area, just as wearing snowshoes does for people. On ice, they walk on the edges of their hooves, which grow out long and sharp for winter. Barren-ground caribou prefer to travel on hardpacked surfaces and commonly track fifty or sixty kilometers across northern lakes during their spring migrations, maintaining a speed of seven or eight kilometers an hour.

When pressed by predators, caribou may hit seventy or eighty kilometers per hour in a wild, hoof-flailing gallop, though they cannot maintain this pace for long. If the attackers are wolves, a short burst of speed is generally enough to discourage further pursuit. Wolves show little interest in mature, healthy caribou; they prey more successfully on the old, the weak, and the very young. Though caribou fawns are remarkably capable — at one day of age they can already outrun a man — mortality is nonetheless high. On the barrens, between forty and eighty percent are likely to die during the first year. Many are lost to storms and disease, but significant numbers of newborns are also taken by wolves.

Once they've survived the perils of youth, caribou find their best defense in numbers. At least, this is true on the open tundra. Forest-dwelling caribou show no inclination to form huge herds. This pattern — large aggregations in open country and a more solitary habit of life in the woods — is typical of the cloven-hoofed mammals. In the forests, an animal can always head for cover when it's frightened. Out on the barrens, a caribou has no place to hide if it's alone, but put it in a herd, and it can take refuge amongst its fellows. The more caribou that are banded together, the less chance there is of a given individual being singled out, and caribou seem to sense this in some way. Often skittish and unapproachable when alone, they may become stolidly indifferent to all but the most immediate danger when they congregate. One biologist tells of getting so close to an unwary stag that he was able to swat it with his rifle butt! If a wolf joins a grazing herd, only the animals that are directly threatened will show real alarm.

Human observers and hunters are not usually perceived as hazards unless they move. And even the animals' suspicion of movement can, at times, be circumvented through a simple ruse. Imagine a person bent at the waist so that his back parallels the ground, with one hand held up to simulate caribou antlers and the other extended downward to represent the neck and head of a grazing animal. It may not look like much of a performance to human eyes, but it is sometimes good enough to fool a caribou.

# Pronghorn

As mentioned earlier, the pronghorn, *Antilocapra americana,* is unique among the modern cloven-hoofed mammals in having horns that are shed annually throughout the animals' lives. In early winter, the outer sheath of hard, fibrous material is cast off, revealing permanent bony cores underneath. This peculiarity places the pronghorn midway between the bovids, with their permanent horns, and the deer, with their deciduous antlers of bone. All the other members of the pronghorn family, including varieties with strangely forked or twisted horn cores, died out in prehistoric times.

The modern pronghorn very nearly followed its vanished relatives down the path to oblivion. Millions of these graceful animals once roamed the great central plains, mingling harmoniously with the bison and foraging for a variety of plants, principally sagebrush and pasture sage, that the bison largely ignored. But by 1908, in response to hunting pressures and the effects of a severe winter in 1906-07, a scattered 20,000 individuals were all that remained of the former multitudes. Today, thanks to decades of management, their population has been restored to the point where limited hunting is allowed.

The speed record for North American mammals is held by the pronghorn. Its playful practice of racing highway vehicles has allowed accurate measurements which indicate that it can sustain speeds of 40 to 60 kilometers per hour for several kilometers. Exceptional individuals may briefly attain a top speed of nearly 100 kilometers per hour. In part, such phenomenal outputs of energy are made possible by its unusually large windpipe and its habit of running open-mouthed.

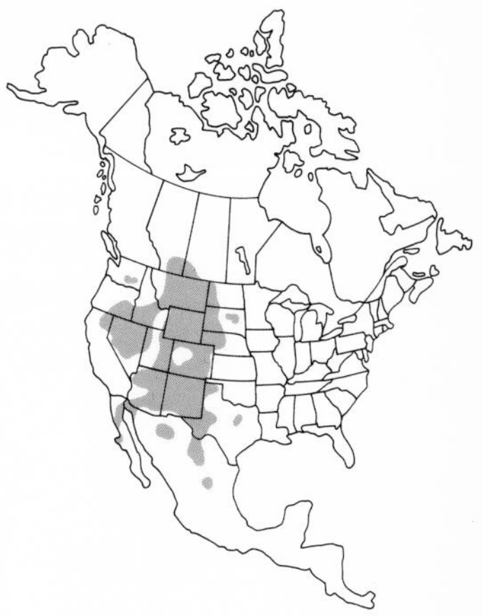

Pronghorn

The pronghorn's fleetness of foot and its almost telescopic eyesight are important elements in its arsenal of defenses against its major predator, the coyote. The habit of forming herds is also part of this defensive strategy. One of the advantages of herding together is that collectively the animals are more likely to spot an approaching predator. If one of them takes alarm and bounds off, its white rump patch fans out and flashes a conspicuous warning signal to other members of the herd.

Although confident runners, pronghorns are strangely unsure of their ability to jump over obstacles. They commonly approach wire-strand fences at a pace that seems nothing less than suicidal, only to dash under or through the wires without a perceptible break in stride. Confronted by a fence that they cannot wiggle or leap through, only the rarest of individuals will jump over. During severe winter storms, when pronghorn herds may migrate southward to escape local conditions, such fences can be deadly obstacles.

The winter herds frequently number as many as a hundred animals of all age classes, but come spring, the sexes tend to segregate into bands of five to ten individuals. Mature females and their yearlings join together in doe herds while most of the sexually mature males form small "bachelor herds." Sparring is a favorite pastime for pairs of bachelors, who dash at one another from a distance of a meter or more to initiate brief shoving matches. The opponent's thrusts are usually caught on the prongs of the horns, which, like deer antlers, appear to have evolved for just this purpose. These minor scuffles establish a dominance rank in which subordinates give way in encounters by moving off or by being chased, butted, or mounted. Dominance ranking is established in the doe herds as well, with access to forage and favorite bedding areas being determined by social position. The bachelors, however, do not reap similar benefits from their rank and persist in squabbling over dominance as an end in itself. Ultimately, some spoils may go to the victor, when at the age of three or four years, bucks that have achieved dominance in the bachelor herds attempt to graduate from their male-only clubs. If successful, they may assume the highest status available to a male pronghorn, that of a territorial buck.

Such an individual imperiously commands

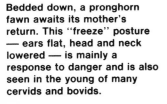

Bedded down, a pronghorn fawn awaits its mother's return. This "freeze" posture — ears flat, head and neck lowered — is mainly a response to danger and is also seen in the young of many cervids and bovids.

Female pronghorns generally have smaller horns than males like the one shown here. The striking pattern of the head and neck markings serves not for camouflage but for territorial and sexual displays.

In winter, pronghorns shed the outer sheath from their horns. The unbranched horn cores show most clearly on the animal that has its head lowered.

a territory of up to four square kilometers for all but the winter months. Territorial boundaries are regularly patrolled and marked with urine, feces, and with a secretion from a special gland in the cheek. Any male interloper is approached and challenged with a loud snort and a peculiar sneeze-like cough. Pressing the encounter with caution, each buck attempts to intimidate his adversary by depositing scent marks or by engaging in ritualistic broadside displays. During the fall rutting season, determined opponents may resort to sparring, which invariably leads to a vicious fight. Injuries may be serious and occasionally cause the death of one of the duelists.

Females that wander into a territorial buck's domain are extended a much different welcome. The buck attempts to herd as many females as possible inside his territory, particularly during the rutting season. Since most mating takes place within such territories, breeding success in a buck reflects the vigor with which he defends a territory and confines his reluctant harem within it.

The does seek solitude for the birth of twin fawns every spring. For their first three weeks of life, the fawns spend most of their time away from the doe, cached in a patch of short vegetation. The doe returns every several hours to nurse the fawns and to watch them relocate to new hiding places. She also consumes their urine and feces, which otherwise would broadcast telltale scents. To foil a potential predator, the doe also exhibits a fierce loyalty toward her newborn offspring. The distressed bleating of a fawn or the sight of a coyote brings the mother bounding to the site of the disturbance. If she cannot lure the coyote away, she may drive it off by charging it and lashing out with her sharp hooves. And it is not only the mother's instincts that are aroused by the beseeching wail of an alarmed pronghorn fawn. All other pronghorns within hearing distance come to investigate, and even related species may be attracted. Imagine the consternation of one biologist who was tagging a crying fawn when 120 huge bison came to observe from a distance of a mere twenty-five meters!

By the time the fawns are three weeks old, their mother's defensiveness has waned and the youngsters join company with other fawns in "nursery herds," usually in the company of some of the does. Romping and racing about in exuberant play, the fawns soon develop dominance relationships which, in females, may determine their lifelong status within the doe herd.

# Bison

The largest land mammal native to Canada is the American buffalo, or, as it is more correctly called, the bison, *Bison bison*. The mature bulls usually stand about 1¾ meters high at the shoulder hump and weigh approximately 900 kilograms, while the cows are shorter and only about half the weight. Both sexes have a dense, shaggy mane that gives the head a disproportionately large appearance, particularly in the summertime, when much of the hair is shed from the hindquarters. Short, black horns, in both males and females, add to their formidable looks.

There are two subspecies of bison, the larger, darker-colored wood bison, which is native to the northern parklands and coniferous forests, and the plains bison, which formerly roamed throughout the central plains of North America. The plains bison were by far the more common and often massed together by the thousands.

These enormous herds were aggregations of many smaller bands, for the basic unit of bison social organization is a herd of five to forty individuals, led by an old cow. The mature bulls come and go from these matriarchal bands and wander alone or in small groups for much of the year. They join the

**Distribution:** Today, bison occur only in wildlife refuges, zoos, and privately owned herds.

cows for longer periods during the rutting season, which occurs in late summer and autumn. Then, for a time, the members of the herd are restless and quarrelsome as the bulls establish dominance relationships or vie for access to estrous females. Disputes between rival bulls may be settled by an exchange of threats — snorting, pawing at the ground, and bellowing ominously, for example. If a serious fight ensues, the animals ram their heads together, but this does little damage since the heads of the males are protected by a thick cushion of hair. They also hook at each other's heads with their sharp horns, sometimes with such powerful thrusts that they lift their opponent's forefeet from the ground. Usually they show a gentlemanly inhibition against pressing a flank attack, but occasionally the rules of battle are discarded, and one of the animals dies from a pierced rib cage or a gored abdomen. Often, a skirmish between two bulls triggers an explosion of other fights throughout the herd as other bulls nervously react to the hostilities.

Another indication of the bison's skittish nature is the ease with which the herds can be frightened into stampeding. Several tribes of Plains Indians, who depended on the bison for food and hides, knew well how to exploit this trait and would send the bison thundering over cliffs or into swamps where they could easily be butchered. One technique commonly used, especially before the introduction of the horse, was to set prairie fires to panic the herds. These fires were in other respects of great benefit to the bison for they helped to maintain the central plains of North America as grassland, preventing trees from encroaching on the bison's habitat.

The prairie grasses once sustained millions upon millions of bison. The early explorers, on ascending a hill, could sometimes view

Facing page: "Buffalo birds," such as cowbirds and blackbirds, do not usually feed on parasites in their hosts' fur. Instead, like the magpie shown here, they search for insects that the bison scares up from the ground as it moves about.

Bison find dust wallows especially inviting in midsummer, when their coats are thin and the flies are at their worst. In the mating season, the bulls also wallow to threaten rivals.

hundreds of thousands at a time, stretching in vast living rivers as far as the eye could see. Food was abundant and the only natural predators were plains grizzlies and packs of wolves, which took mainly the young, the sick, and the very old. But the predator that ultimately decimated the thriving population of the last century was the white man.

The virtual extermination of the plains bison took place in the 1870s and 1880s, at the hands of buffalo hunters equipped with guns and horses. Some 5,000 hunters and skinners plied their grisly trades on the prairies, annually sending as many as 200,000 buffalo robes to eastern markets. So immense was the destruction that the scattered bones of the bison were for years an important source of income to settlers, who gathered them to be sent east for the manufacture of fertilizer. It is estimated that from Saskatoon alone, trains carried away 3,000 carloads of bones. By the mid-1880s a few strays and a few small domestic herds, a total of several hundred animals, were all that remained of the multitudes that once had darkened the plains.

The cause of conserving and protecting the remaining bison was eventually taken up by the Canadian government. A domestic herd of plains bison, purchased in 1906, formed the basis of the largest present-day Canadian herds, those found in Elk Island and Wood Buffalo National Parks. The population of the other subspecies of bison, the wood bison, had shrunk to about 500 when, in 1893, the Canadian government passed legislation to protect it. When Wood Buffalo National Park was established in 1922, the numbers of wood bison had recovered to about 2,000. Today, the park contains a population of about 14,000, consisting of crosses between wood bison and the plains bison that were introduced there in the 1920s. A remnant herd of the woodland subspecies was discovered in the park in 1957. From this herd, individuals with fairly pure wood bison characteristics were selected and relocated to the Mackenzie Bison Sanctuary and to Elk Island National Park.

Although the former multitudes of wild bison have vanished forever from the plains, some trace of them may still be discerned today. On patches of unbroken prairie, you may see certain shallow depressions overgrown with vegetation. These are wallows, which resulted from the bison's habit of rolling in the dust or mud as a remedy against the summer flies.

# Mountain Goat

Mountain goat

Mountain goats, *Oreamnos americanus,* are well adapted to survive on the most rugged mountain heights. Their shaggy white coats repel the icy mountain winds, while their hooves are specially equipped with nonskid soles and highly flexible toes for conquering the dizzying heights of rock and ice. These dauntless mountaineers may ascend to the very peaks of the mountains, foraging for grass on the ledges and steep slopes and bedding down at night in the most inaccessible areas.

As winter approaches, the descending snow line may prompt them to seek food at lower and lower elevations. Another of their strategies in winter, for avoiding deep snow, is to seek out cliffs and windswept ridges at higher altitudes. In either case, they sometimes form herds: if the snow in most places is deep, they may converge on whatever large patches of winter browse are available. The largest groups, which include five or ten individuals or more, usually consist of females with young; the males are less sociable and wander alone or in small groups. Mixed herds are more likely to be formed in the summertime, around salt licks.

For goats that band together, the social order that prevails in most other societies of cloven-hoofed mammals is reversed — except during the rut, the females, or nannies, especially those with kids, are usually dominant over any males, or billies, that happen to be

**Mountain goats are less sociable than many other ungulates, but the bond between a nanny and her kid is strong.**

present. This is surprising because the billies are larger and more powerful than the nannies and are armed with the same formidable weapons — short, bayonet-like horns. But there is an important benefit to this arrangement: the nannies exploit their dominant status by monopolizing the best feeding and bedding areas, so even after the harshest of winters some of them survive to perpetuate the species with a new crop of kids.

The kids are born around the first of June, usually on some remote ledge on the face of a cliff. One kid per female is most common, but a nanny may also bear twins or triplets. The mother is extremely protective for the first year of her kid's life, feeding and bedding with it, walking on its down-slope side to guard it from falling, and sharing with it the winter feeding craters that she paws in the snow. In the summer, when the nannies and their kids congregate in nursery herds, the kids romp and play together wildly, circling and butting each other, playing "king of the castle" on a ledge or rock, even leaping over each other or their mothers. The oddest feat performed by these enthusiastic acrobats is "whirling," a stunt usually performed on a slope and consisting of leaping into the air, then twisting horizontally, often through one or more complete turns. Occasionally this youthful silliness infects an entire band, even the normally sedate adults, and they all engage in an undignified bout of whirling as they descend a slope.

The strong bond between a nanny and her kid is the basis of social organization in mountain goats. The yearling follows its nanny until she drives it off just before the birth of her new kid. After this separation, the sexually immature males (those up to two years of age) and females of all ages often find another female with a kid to follow. The mature males often visit these maternal bands during the rutting season, in November.

When two billies dispute the right to court a female, the result may be a vicious and bloody fight. Circling each other head to tail, each stabs at his opponent's belly, flank, and rear with his daggerlike horns, inflicting deep, often fatal wounds. However, the billies are not totally unprotected from these assaults, for nature has equipped them with a type of shield. Their skins are so thick — five or six millimeters on the hindquarters and up to twenty-two millimeters on the rumps of some very old males — that the Alaska Coastal Indians were able to use the hides of mountain goats for breast armor. The extremely damaging broadside-fighting technique of mountain goats is probably a primitive style of combat that was used by the ancestors of all our native cloven-hoofed mammals. Most modern-day relatives of the mountain goat have evolved safer techniques for fighting: head-to-head wrestling, as in deer and pronghorns; or head-on ramming, as in mountain sheep.

Not surprisingly, mountain goats are extremely reluctant to fight and prefer a bluffing match to outright battle. They may rush at one another, stopping short of actual contact and raking the air with their horns. Or they may slowly circle each other, humming ominously and occasionally roaring, all the while arching their backs to increase their apparent size. This explains why they have a crest of hair on the neck and back — it evolved to make them look larger during these broadside displays. Sometimes they sit down on their haunches, like dogs, and furiously paw "rutting pits" in the ground with a forefoot, blackening their rumps, bellies, and legs with earth.

At the beginning of the rutting season, the nannies disdain the advances of the billies and repel their suitors with threats or outright horn attacks. Only at the height of the rut are the males given any quarter, and even then their approach to the females is a humble, groveling crawl. This crouching, head-on approach is designed to appease the female by presenting an appearance that is quite opposite to the broadside threat display.

The lethal horns of mountain goats are sometimes directed at predators such as cougars, eagles, lynx, coyotes, and wolverines. Even a grizzly bear can be driven off by a counterattack from an enraged mountain goat. But predators are seldom a concern of the nimble-footed goat, unassailable in its lofty fortress of cliffs and snow. And yet the moun-

**Even at the top of its icy world, the mountain goat is secure against the cold in its long winter coat.**

tains are an untrustworthy protector, for snowslides and rockslides claim large numbers of victims, and many goats bear the marks of a misplaced step — broken horns or lame legs.

The greatest threat to the survival of mountain goats is mankind. When mining and forestry roads penetrate to the heart of mountain goat country, hunters quickly decimate the local goat populations, taking trophy heads and usually leaving the meat to scavengers. Mountain goats cannot sustain heavy hunting pressures, and their Canadian population declined from about 100,000 in 1964 to some 40,000 in the late 1970s.

# Muskox

**Muskox**

Some of the bleakest, harshest terrain in Canada is found on the tundra of the northern mainland and arctic islands. Here, for most of the year, the low-lying vegetation is frozen beneath the snow. And at the latitude of Ellesmere Island, in the high arctic, the sun fails to rise for four months, between November and February. But this environment, inhospitable as it is to people, is nevertheless the year-round home to Canada's population of muskoxen, *Ovibos moschatus.* These magnificent and stately beasts have many special adaptations that uniquely suit them to survive the gruelling winters and to prosper during the arctic's brief summers.

Even the bitterest of cold seldom affects these animals. A fine, cashmere-like undercoat provides insulation, while long guard hairs, hanging nearly to the ground, provide a heavy overcoat. The large body size (an average of 350 kilograms for a bull and 300 kilograms for a cow), the diminutive tail, and the relatively short, stocky legs help the animals to retain their body heat by reducing their surface-to-volume ratio.

Cold is seldom a problem even for the calves. Although most of them are born in May when snow cover is deepest and temperatures can still drop to minus thirty degrees Celsius, the newborns are usually protected adequately by their short, curly fur. By being born very early in the spring, they are able to take full advantage of the lush grazing that summer provides. When only a week or two old, the calves begin to supplement their intake of milk with the adult diet of Labrador tea, willows, grasses, sedges, and other plants. Calves and adults alike fatten rapidly when the new vegetation appears in mid-June.

In the relatively hot summers, overheating can be a problem for an animal dressed in a five-centimeter layer of wool; accordingly, it is at this time that the adults renew their old coats. The underhair is shed in May or June, and clumps hanging from the longer, outer hairs give the animals a shaggy, moth-eaten appearance that may persist through the summer.

The hooves of muskoxen, like those of other cloven-hoofed mammals, are designed to meet special needs imposed on the animals by their environment. A soft inner heel, similar to that found in mountain goats, provides traction on smooth, rocky surfaces, while a hard, sharp, outer edge enables swift travel over hard-packed snow. The front hooves are larger than the rear ones, an advantage when it comes to pawing the snow away from winter feeding craters. Deep snow cover isn't often a problem, for in the winter the muskoxen generally feed in areas with shallow snow, either in the lowlands, or on hills, where the wind usually reduces snow depth to a few centimeters. But when snowfall is heavy or when freezing rains lock up the larder of vegetation, mortality rates soar, and the special adaptations of the hooves are put to the test in the fight for survival.

In addition to structural specializations, the muskoxen also have behavioral adaptations to the tundra environment. For example, the calves may supplement their grazing by

**The Inuit call the muskox *omingmak* — "the bearded one." The union of the horn bosses across the forehead indicates that this animal is a mature bull.**

Perceiving a threat to their safety, a herd of muskoxen closes ranks for defense.

suckling until the unusual age of a year or more, a trick that helps them through periods of environmental stress. Muskoxen are also adapted to the barrens by their sociability, which unites them into herds of ten or twenty individuals in the summer and up to sixty in the winter. On the treeless tundra, where vegetation offers no concealment, the muskoxen are easily spotted by wolves, their chief natural predators. When threatened by wolves, the herd usually runs to high ground or to an area of shallow snow and closes ranks in a straight line, horns directed towards the attackers. When surrounded, the herd forms a tight defensive circle, with the adults facing outwards and the calves and yearlings pressed close to their mothers' sides. If a wolf approaches closely, an adult, often the dominant bull, may attempt to hook it with its sharp horns or may make a short charge, backing into the defensive unit once again when the foray is over. Occasionally, one of the muskoxen may break ranks and attempt to flee, only to be driven back into the formation by an attack from the lead bull. Wolves seldom penetrate this defense, but they do frequently kill lone bulls or stragglers that don't group together quickly enough.

Effective as it is against wolves, the muskoxen's defensive formation is only a liability in the face of hunters equipped with firearms. Muskoxen were once slaughtered to the edge of extinction by hide hunters and were saved only at the last minute by government protection, which was begun on the mainland in 1917 and extended to the arctic islands in 1926. Muskox numbers increase slowly, because the cows bear their first calves at the relatively late age of four years and don't always produce a calf every year thereafter. But like the bison, they have responded to government management, and today more than 10,000 muskoxen inhabit the Canadian arctic.

While muskoxen are similar in appearance to bison, anatomical evidence shows them to be more closely related to sheep and goats. This should come as no surprise to anyone who has witnessed muskoxen bulls fighting for the possession of harems during the August rutting period. Like mountain sheep, they begin their attacks by charging each other from a considerable distance, the thick bosses at the base of the horns meeting in a bone-jarring collision. And if jousting in this fashion doesn't decide the winner, they will, like fighting mountain goats, attempt to gore the flanks of their opponents with their sharp horns, occasionally with fatal consequences.

# Mountain Sheep

Two species of mountain sheep are native to the foothills, mountain pastures, and alpine meadows of western North America: the bighorn sheep, *Ovis canadensis,* and its smaller relative, the thinhorn sheep, *Ovis dalli.* These species could potentially interbreed, but they were separated almost 20,000 years ago and have remained apart ever since. Their common ancestor, like the ancestors of most of North America's other cloven-hoofed mammals, emigrated from Asia across the Bering land bridge. Then, during the last ice age, the irresistible advance of the glaciers drove this primitive population of sheep into two ice-free areas, one in Alaska and the Yukon, the other south of a great ice sheet that spanned the width of the continent. The southern survivors evolved into the bighorns and repopulated western areas of Canada and the United States as the glaciers slowly withdrew. Sheep from the northern refuge pursued a slightly different evolutionary course to become thinhorn sheep.

In Canada, there are two races of thinhorn sheep, distinguished from one another by their color. A member of the northern race, Dall's sheep, is predominantly white, while one from the southern race, Stone's sheep, looks rather as if a dark gray overcoat had been thrown

**Facing page: For most of the year, bands of bighorn ewes, such as the one shown here, contain only adult females and their sexually immature offspring. The short horns are characteristic of adult ewes and yearling rams.**

over it, leaving its white head, rump, and belly exposed. Since these two races are not geographically isolated, interbreeding produces a hybrid of mixed appearance where the populations merge.

Bighorn and thinhorn sheep are both highly gregarious and guide their social behavior by similar rules. Conservative in their habits, groups of sheep occupy a series of home ranges, returning to the same area at the same season, year after year. In the depth of winter, when the mountain passes and alpine meadows are choked with snow, the sheep are to be found on lower ranges, where they are able to paw through the snow for the grasses, sedges, and forbs that make up their diet. If strong winds sweep the snow from cliffs and ridges, then the animals extend their winter range to higher elevations, where foraging is easier. As spring lifts the blanket of snow from the mountains, the sheep follow the retreating edge upwards, nibbling on the sprouting vegetation as they go.

The pregnant ewes then seek solitude on separate ranges for the birth of their lambs. Situated in rough terrain with precipitous cliffs, the lambing ranges provide the surefooted ewe and her newborn lamb with the best chance of escape from their chief enemies, wolves and cougars. At first the lamb stays close to its mother, gaining protection from predators such as eagles. If a soaring eagle swoops to attack, the lamb scurries under its mother's belly while she repels the assault with her horns. But by the time the lamb is two weeks old, it prefers to frolic with other lambs, returning to its dam only to nurse. The loose bond between ewe and lamb benefits the species when a wolf or a cougar threatens the lamb: since the ewe is no match for these

**The horns of this bighorn ram tell a story: the broken tip was probably sustained in a battle with a rival; and the more prominent rings, one for each year of growth, indicate an age of more than eight years.**

predators, she escapes and saves herself, perhaps to perpetuate her kind another year. It may seem strange to us that the ewe lacks the maternal loyalty so fiercely displayed by other cloven-hoofed mammals, such as moose or pronghorns. But remember that a healthy moose is powerful enough to repel wolves, and pronghorns usually only have to contend with coyotes.

A few weeks after lambing, the ewes and their lambs, along with the sexually immature offspring from previous seasons, band together in groups of ten or more under the leadership of an older, experienced ewe, usually one with a lamb. Throughout the summer, they graze placidly in alpine meadows, relying on their acute eyesight to warn of approaching danger in this treeless habitat.

The sexually mature rams, those two or three years of age and older, form their own bands of five or ten individuals (rarely as many as fifty) and occupy spring and summer ranges separate from those of the ewe groups. Like the ewes, the rams follow a dominant individual as they move between ranges. They remain together until the rutting season in late November and December when they trek as far as sixty kilometers to join the ewes on the lower, winter ranges.

It is during the rut that the males stage their most spectacular battles. Following an ancient, hereditary ritual, two antagonists rear up on their hind legs, dash towards each other, and throw their weight into a horn-to-horn collision. As the crash reverberates through the mountain valleys, they freeze like statues for a moment, displaying their horns to one another. The contest occasionally ends after a single clash, but if the adversaries are evenly matched, they may joust on and off for an

Compared to bighorn rams, the Dall's sheep rams, shown here, have thinner, more widely flared horns.

A Dall's sheep lamb subjects its mother to one of the tribulations of parenthood. On the face of this steep cliff, the lamb is safe from most predators, except for eagles or perhaps a mountain lion.

Thinhorn sheep
Bighorn sheep

entire day and night until one or both succumb to exhaustion.

The combatants avoid serious injury not by any built-in inhibition against intraspecific aggression, such as is found in bison or mountain goats, but rather by their superb defenses. The blows are usually caught with the horns, but the head is further armored by two roofs of bone, with a honeycomb of bone reinforcement in between. Strong ligaments support the skull, and, in a collision, the head rotates forward, allowing the impact to be partially absorbed by a thick, elastic tendon running from the skull over the top of the spine. The horns of the rams grow throughout life, completing a full curl after seven or eight years. (In contrast, the horns of the females are only curved spikes.) At maturity, the massive skull and horns of a ram weigh about twelve kilograms (exceptionally fifteen) for a bighorn, less for a thinhorn ram.

Since rams fight primarily for dominance over one another, rather than for access to females, clashes occur throughout most of the year, not only during the fall rut. On their summer ranges, less serious sparring amongst the rams establishes an order of dominance that is based on horn size. During an altercation, a subordinate neutralizes the aggressiveness of a dominant ram by behaving like a female, sometimes displaying his white rump patch and permitting the dominant to mount him. In ram groups, therefore, each individual assumes a female role with superiors and a male role with subordinates.

Energy is seldom wasted in hopeless contests between unevenly matched rams, since each ram can judge the probable outcome of a quarrel from the size of his adversary's horns. Ranking by horn size thus establishes a degree of harmony that permits the highly aggressive rams to live together throughout the spring and summer. Severe fights usually occur only on the rutting grounds when two strangers of nearly equal horn size meet and resort to combat to settle the question of superiority.

The victorious ram wins the right to consort with any ewe in estrus. He isn't always successful in thwarting the mating attempts of subordinate rams, however, and the constant attempt to defend females depletes fat reserves that the ram badly needs for the winter months. Weakened by the fall rut and ensuing winter food scarcity, he is more likely to succumb to a predator. Amongst bighorns, for example, vigorous males with large horns die off rapidly after reaching full maturity at eight or nine years and seldom live past fifteen years. On the other hand, smaller-horned, socially unsuccessful males may reach nineteen years of age, and females may live to be twenty-four years old. Thus the dominant ram pays a price for his social and reproductive success.

# Checklist

The following 163 species of wild mammals are native to western Canada. Those marked with a double asterisk are illustrated and/or discussed in detail in this book. Those marked with a single asterisk are treated more briefly.

**Marsupials**

*Opossum, or Virginia opossum, *Didelphis virginiana*

**Insectivores**

**Masked, or cinereous, shrew, *Sorex cinereus*
Vagrant shrew, *Sorex vagrans*
Dusky shrew, *Sorex monticolus*
Water, or northern water, shrew, *Sorex palustris*
Pacific water, or Bendire's, shrew, *Sorex bendirii*
Arctic shrew, *Sorex arcticus*
Trowbridge's shrew, *Sorex trowbridgii*
*Pygmy shrew, *Microsorex hoyi*
**Short-tailed shrew, *Blarina brevicauda*
Shrew-mole, *Neurotrichus gibbsii*
Townsend's mole, *Scapanus townsendii*
Coast, or Pacific coast, mole, *Scapanus orarius*
*Star-nosed mole, *Condylura cristata*

**Bats**

**Little brown bat, *Myotis lucifugus*
Yuma bat, *Myotis yumanensis*
Keen's bat, *Myotis keenii*
Northern long-eared bat, *Myotis septentrionalis*
Pale long-eared bat, *Myotis evotis*
Fringed bat, *Myotis thysanodes*
Long-legged bat, *Myotis volans*
California bat, *Myotis californicus*
Small-footed bat, *Myotis leibii*
Silver-haired bat, *Lasionycteris noctivagans*
Big brown bat, *Eptesicus fuscus*
Red bat, *Lasiurus borealis*
**Hoary bat, *Lasiurus cinereus*
Spotted bat, *Euderma maculatum*
Townsend's, or western, big-eared bat, *Plecotus townsendii*
Pallid bat, *Antrozous pallidus*

**Lagomorphs**

**Collared pika, *Ochotona collaris*
**Rocky Mountain pika, *Ochotona princeps*
**Eastern cottontail, *Sylvilagus floridanus*
**Nuttall's, or mountain, cottontail, *Sylvilagus nuttallii*
**Snowshoe, or varying, hare, *Lepus americanus*
**Arctic hare, *Lepus arcticus*
**White-tailed jack rabbit, *Lepus townsendii*

**Rodents**

*Mountain beaver, *Aplodontia rufa*
Eastern chipmunk, *Tamias striatus*
**Least chipmunk, *Eutamias minimus*
**Yellow-pine chipmunk, *Eutamias amoenus*
Townsend's chipmunk, *Eutamias townsendii*
Red-tailed chipmunk, *Eutamias ruficaudus*
**Woodchuck, or groundhog, *Marmota monax*
**Yellow-bellied marmot, *Marmota flaviventris*
**Hoary marmot, *Marmota caligata*
*Vancouver Island, or Vancouver, marmot, *Marmota vancouverensis*
**Richardson's ground squirrel, *Spermophilus richardsonii*
**Columbian ground squirrel, *Spermophilus columbianus*
**Arctic ground squirrel, *Spermophilus parryii*
**Thirteen-lined ground squirrel, *Spermophilus tridecemlineatus*
*Franklin's ground squirrel, or gray gopher, *Spermophilus franklinii*
**Golden-mantled ground squirrel, *Spermophilus lateralis*
*Cascade golden-mantled ground squirrel, *Spermophilus saturatus*
Black-tailed prairie dog, *Cynomys ludovincianus*
Gray or black squirrel, *Sciurus carolinensis*
**Red squirrel, *Tamiasciurus hudsonicus*
*Douglas's squirrel, *Tamiasciurus douglasii*
**Northern flying squirrel, *Glaucomys sabrinus*
Northern pocket gopher, *Thomomys talpoides*
Plains pocket gopher, *Geomys bursarius*
Olive-backed pocket mouse, *Perognathus fasciatus*
Great Basin pocket mouse, *Perognathus parvus*
Ord's kangaroo rat, *Dipodomys ordii*
**Beaver, *Castor canadensis*
Western harvest mouse, *Reithrodontomys megalotis*
**Deer mouse, *Peromyscus maniculatus*
Sitka mouse, *Peromyscus sitkensis*
White-footed mouse, *Peromyscus leucopus*
Northern grasshopper mouse, *Onchomys leucogaster*
**Bushy-tailed wood rat, *Neotoma cinerea*
Northern red-backed vole, *Clethrionomys rutilus*
*Gapper's, or southern, red-backed vole, *Clethrionomys gapperi*
Western red-backed vole, *Clethrionomys occidentalis*
Heather vole, *Phenacomys intermedius*
**Meadow vole, *Microtus pennsylvanicus*

Montane vole, *Microtus montanus*
Townsend's vole, *Microtus townsendii*
Tundra, or root, vole, *Microtus oeconomus*
Long-tailed vole, *Microtus longicaudus*
Yellow-cheeked, or chestnut-cheeked, vole, *Microtus xanthognathus*
Creeping vole, *Microtus oregoni*
Singing vole, *Microtus gregalis*
Prairie vole, *Microtus ochrogaster*
Water, or Richardson's water, vole, *Arvicola richardsonii*
Sagebrush vole, *Lagurus curtatus*
**Muskrat, *Ondatra zibethicus*
Brown lemming, *Lemmus sibiricus*
Southern bog lemming, *Synaptomys cooperi*
Northern bog lemming, *Synaptomys borealis*
Collared lemming, *Dicrostonyx torquatus*
Meadow jumping mouse, *Zapus hudsonius*
Western jumping mouse, *Zapus princeps*
Pacific jumping mouse, *Zapus trinotatus*
Woodland jumping mouse, *Napaeozapus insignis*
**Porcupine, *Erethizon dorsatum*

**Toothed whales**

North Pacific bottled-nosed, or Baird's beaked, whale, *Berardius bairdii*
North Pacific, or Stejneger's, beaked whale, *Mesoplodon stejnegeri*
Moore's beaked, or archbeak, whale, *Mesoplodon carlhubbsi*
Goose-beaked whale, *Ziphius cavirostris*
Sperm whale, *Physeter macrocephalus*
Beluga, or white whale, *Delphinapterus leucas*
Narwhal, *Monodon monoceros*
Striped, or blue, dolphin, *Stenella coeruleoalba*
Northern right-whale dolphin, *Lissodelphis borealis*
*Pacific white-sided dolphin, *Lagenorhynchus obliquidens*
**Killer whale, *Orcinus orca*
Gray grampus, or Risso's dolphin, *Grampus griseus*
Short-finned, or Pacific, pilot whale, *Globicephala macrorhynchus*
**Harbor porpoise, *Phocoena phocoena*
**Dall's porpoise, *Phocoenoides dalli*

**Baleen whales**

**Gray whale, *Eschrichtius robustus*
Fin whale, *Balaenoptera physalus*
Sei whale, *Balaenoptera borealis*
Little piked, or minke, whale, *Balaenoptera acutorostrata*
Blue, or sulphur-bottomed, whale, *Balaenoptera musculus*
**Humpback whale, *Megaptera novaeangliae*
Black right, or right, whale, *Balaena glacialis*
Bowhead whale, *Balaena mysticetus*

**Carnivores**

**Coyote, *Canis latrans*
**Wolf, or gray wolf, *Canis lupus*
**Arctic fox, *Alopex lagopus*
**Red fox, *Vulpes vulpes*
Gray fox, *Urocyon cinereoargenteus*
**Black bear, *Ursus americanus*
**Grizzly, or brown, bear, *Ursus arctos*
**Polar bear, *Ursus maritimus*
**Raccoon, *Procyon lotor*
**Marten, *Martes americana*
**Fisher, *Martes pennanti*
**Ermine, or stoat, or short-tailed weasel, *Mustela erminea*
**Least weasel, *Mustela nivalis*
**Long-tailed weasel, *Mustela frenata*
**Mink, *Mustela vison*
**Wolverine, *Gulo gulo*
**Badger, *Taxidea taxus*
**Western spotted skunk, *Spilogale gracilis*
**Striped skunk, *Mephitis mephitis*
**River otter, *Lutra canadensis*
**Sea otter, *Enhydra lutris*
**Mountain lion, or cougar, *Felis concolor*
**Lynx, *Felis lynx*
**Bobcat, *Felis rufus*

**Pinnipeds**

*Northern fur seal, *Callorhinus ursinus*
**Steller, or northern, sea lion, *Eumetopias jubatus*
*California sea lion, *Zalophus californianus*
*Walrus, *Odobenus rosmarus*
**Harbor, or common, seal, *Phoca vitulina*
*Ringed seal, *Phoca hispida*
Harp seal, *Phoca groenlandica*
Bearded seal, *Erignathus barbatus*
*Northern elephant seal, *Mirounga angustirostris*

**Cloven-hoofed mammals**

**Wapiti, or elk, *Cervus elaphus*
**Mule, or blacktail, deer, *Odocoileus hemionus*
**White-tailed deer, *Odocoileus virginianus*
**Moose, *Alces alces*
**Caribou, *Rangifer tarandus*
**Pronghorn, *Antilocapra americana*
**Bison, or buffalo, *Bison bison*
**Mountain goat, *Oreamnos americanus*
**Muskox, *Ovibos moschatus*
**Bighorn sheep, *Ovis canadensis*
**Thinhorn sheep, *Ovis dalli*

# References

This is by no means a complete bibliography. Only those sources that contributed information to the text or photo captions are included here. Major sources are marked with an asterisk.

## General references

Allen, T. B., ed. 1979. *Wild animals of North America.* Washington, D.C.: National Geographic Society.

Anderson, S., and Jones, J. K., Jr., eds. 1967. *Recent mammals of the world: a synopsis of families.* New York: Ronald Press. This book served as a supplementary reference on mammalian taxonomy down to and including the levels of the family and subfamily.

Banfield, A. W. F. 1974. *The mammals of Canada.* Toronto: University of Toronto Press. Although the text is frequently inaccurate, this book contains good distribution maps, which were amongst the references used in preparing maps for the present volume. It was also used as a supplementary source of vernacular names for the text and checklist.

Burt, W. H., and Grossenheider, R. P. 1976. *A field guide to the mammals.* 3rd ed. Boston: Houghton Mifflin.

Cowan, I. McT., and Guiguet, C. J. 1956. *The mammals of British Columbia.* Victoria: British Columbia Provincial Museum.

Deems, E. F., Jr., and Pursley, D., eds. 1978. *North American furbearers: their management, research and harvest status in 1976.* Publ. Fur Resources Committee, International Association of Fish and Wildlife Agencies. University of Maryland Press, College Park. The detailed distribution maps in this work were very useful in preparing some of the maps in the present volume.

Hall, E. R., and Kelson, K. R. 1959. *The mammals of North America.* 2 vols. New York: Ronald Press. This work served as a supplementary reference on taxonomy and nomenclature at the species level of classification. It was also one of the references used in preparing distribution maps.

Jones, J. K., Jr.; Carter, D. C.; and Genoways, H. H. 1979. *Revised checklist of North American mammals north of Mexico.* Texas Tech. U. Occ. Paper No. 62. This list served as our main authority on mammalian taxonomy and nomenclature.

Kurtén, B. 1971. *The age of mammals.* London: Weidenfeld and Nicolson. This is a basic reference on mammalian paleontology and evolution.

Lawler, T. E. 1979. *Handbook to the orders and families of living mammals.* 2nd. ed. Eureka, Calif.: Mad River Press. We have followed this book in treating the Pinnipedia as a separate order rather than as part of Carnivora.

Morris, D. 1965. *The mammals: a guide to the living species.* New York: Harper and Row.

Murie, O. J. 1975. *A field guide to animal tracks.* Boston: Houghton Mifflin.

Peterson, R. L. 1966. *Mammals of eastern Canada.* Toronto: Oxford University Press.

Romer, A. W. 1966. *Vertebrate paleontology.* 3rd ed. Chicago: University of Chicago Press.

Seton, E. T. 1929 (Reprint 1953). *Lives of game animals.* 4 vols. Boston: Charles T. Branford. This work is now of interest mainly as a source of historical and anecdotal information.

Soper, J. D. 1964. *The mammals of Alberta.* Edmonton: Alberta Department of Industry and Development.

Vaughan, T. A. 1978. *Mammalogy.* 2nd ed. Philadelphia: W. B. Saunders.

Walker, E. P. 1964. *Mammals of the world.* 3 vols. Baltimore: Johns Hopkins Press.

Youngman, P. M. 1975. *Mammals of the Yukon Territory.* National Museums of Canada Publications in Zoology No. 10. The distribution maps in this book were useful in preparing maps for the present volume.

## Introduction

Canada Dept. of Mines and Technical Surveys, Geographical Branch. 1957. *Atlas of Canada.* Map 38. Ottawa.

Canada Dept. of Energy, Mines and Resources, Surveys and Mapping Branch. 1973. *The national atlas of Canada.* Folio B, leaves 45-46. Ottawa.

Carter, G. F. 1978. The American paleolithic. In *Early man in America from a circum-Pacific perspective,* ed. A. L. Bryan, pp. 10-19. Edmonton: Archaeological Researches International.

Gunderson, H. L. 1976. *Mammalogy.* New York: McGraw-Hill.

Hopkins, D. M., ed. 1967. *The Bering land bridge.* Stanford, Calif.: Stanford University Press.

Jones, J. K., Jr., and Armstrong, D. M. Mammalia. *Encyclopaedia Britannica Macropaedia* 11: 401-16.

Martin, P. S., and Wright, H. E., Jr., eds. 1967. *Pleistocene extinctions: the search for a cause.* New Haven: Yale University Press.

Van Gelder, R. G. 1969. *Biology of mammals.* New York: Charles Scribner's Sons.

## Shrews and moles

Buckner, C. H. 1966. Populations and ecological relationships of shrews in tamarack bogs of southeastern Manitoba. *J. Mamm.* 47: 181-94.

Criddle, S. 1973. The granivorous habits of shrews. *Can. Field-Nat.* 87: 69-70.

Doucet, G. J., and Bidet, J. R. 1974. The effects of weather on the activity of the masked shrew. *J. Mamm.* 55: 348-63.

Eadie, W. R. 1944. The shorttailed shrew and field mouse predation. *J. Mamm.* 25: 359-64.

Findley, J. S. 1974. Insectivora. *Encyclopaedia Britannica Macropaedia* 20: 622-28.

Goodwin, M. K. 1979. Notes on caravan and play behavior in young captive *Sorex cinereus. J. Mamm.* 60: 411-13.

Gould, E.; Negus, N. C.; and Novick, A. 1964. Evidence for echolocation in shrews. *J. Exp. Zool.* 156: 19-38.

Hawes, M. L. 1977. Home range, territoriality, and ecological separation in sympatric shrews, *Sorex vagrans* and *Sorex obscurus. J. Mamm.* 58: 354-67.

Lawrence, B. 1946. Brief comparison of short-tailed shrew and reptile poisons. *J. Mamm.* 26: 393-96.

Pearson, O. P. 1948. Metabolism of small mammals, with remarks on the lower limit of mammalian size. *Science* 108: 44.

Platt, W. J. 1976. The social organization and territoriality of short-tailed shrew (*Blarina brevicauda*) populations in oldfield habitats. *Anim. Behav.* 24: 305-18.

Pruitt, W. O., Jr., 1954. Notes on a litter of young masked shrews. *J. Mamm.* 35: 109-10.

Tomasi, T. E. 1979. Echolocation by the short-tailed shrew *Blarina brevicauda. J. Mamm.* 60: 751-59.

## Bats

Barbour, R. W., and Davis, W. H. 1969. *Bats of America.* Kentucky: University Press.

Dorward, W. J.; Schowalter, D. B.; and Gunson, J. R. 1977. Preliminary studies of bat rabies in Alberta. *Can. Vet. J.* 18: 341-48.

Fenton, M. B. 1974. Echoes: why bats can live in caves. *Can. Geog. J.* 98: 16-23.

———. 1977. Variation in the social calls of little brown bats (*Myotis lucifugus*). *Can. J. Zool.* 55: 1151-57.

———. 1978. Bats. *Nature Can.* 7(3): 32-38.

Fenton, M. B., and Bell, G. P. 1979. Echolocation and feeding behaviour in four species of *Myotis* (Chiroptera). *Can. J. Zool.* 57: 1271-77.

Fenton, M. B.; van Zyll de Jong, C. J.; Bell, G. P.; Campbell, D. P.; and Laplante, M. 1980. Distribution, parturition dates, and feeding of bats in south-central British Columbia. *Can. Field-Nat.* 94: 416-20.

Fullard, J. H.; Fenton, M. B.; and Simmons, J. A. 1979. Jamming bat echolocation: the clicks of arctiid moths. *Can. J. Zool.* 57: 647-49.

Fullard, J. H., and Barclay, R. M. R. 1980. Audition in spring species of arctiid moths as a possible response to differential levels of insectivorous bat predation. *Can. J. Zool.* 58: 1745-50.

Gould, E.; Negus, N. C.; and Novick, A. 1964. Evidence for echolocation in shrews. *J. Exp. Zool.* 156: 19-38.

Griffin, D. R.; Webster, F. A.; and Michael, C. R. 1960. The echolocation of flying insects by bats. *Anim. Behav.* 8: 141-54.

Hitchcock, H. B. 1973. *Bat.* Ottawa: Canadian Wildlife Service Hinterland Who's Who series.

Humphrey, S. R., and Cope, J. B. 1976. *Population ecology of the little brown bat,* Myotis lucifugus, *in Indiana and north-central Kentucky.* American Society of Mammalogists, Special Publication No. 4.

Keen, R., and Hitchcock, H. B. 1980. Survival and longevity of the little brown bat (*Myotis lucifugus*) in southeastern Ontario. *J. Mamm.* 61: 1-7.

Leen, N., and Novick, A. 1969. *The world of bats.* New York: Holt, Rinehart and Winston.

McManus, J. J. 1974. Activity and thermal preference of the little brown bat, *Myotis lucifugus,* during hibernation. *J. Mamm.* 55: 844-46.

Schowalter, D. B. 1980. Swarming, reproduction, and early hibernation of *Myotis lucifugus* and *M. volans* in Alberta, Canada. *J. Mamm.* 61: 350-54.

Schowalter, D. B.; Gunson, J. R.; and Harder, L. D. 1979. Life history characteristics of little brown bats (*Myotis lucifugus*) in Alberta. *Can. Field-Nat.* 93: 243-51.

Thomas, D. W.; Fenton, M. B.; and Barclay, R. M. R. 1979. Social behavior of the little brown bat, *Myotis lucifugus.* 1. Mating behavior. *Behav. Ecol. Sociobiol.* 6: 129-36.

Webster, F. A., and Griffin, D. R. 1962. The role of

the flight membranes in insect capture by bats. *Anim. Behav.* 10: 332-40.

Wimsatt, W. A., ed. 1977. *Biology of bats,* vols. 1 and 3. New York: Academic Press.

Yalden, D. W., and, Morris, P. A. 1975. *The lives of bats.* New York: Quadrangle.

## Introduction to pikas, rabbits, and hares

Dawson, M. R. 1974. Lagomorpha. *Encylopaedia Britannica Macropaedia* 10: 588-91.

Johnson, D. R., and Mazell, M. H. 1966. Energy dynamics of Colorado pikas. *Ecol.* 47: 1059-61.

Orr, R. T. 1977. *The little known pika.* New York: Macmillan.

## Pikas

Barash, D. P. 1973. Territorial and foraging behavior of pika (*Ochotona princeps*) in Montana. *Am. Midl. Nat.* 89: 202-7.

Broodbooks, H. E. 1965. Ecology and distribution of the pikas of Washington and Alaska. *Am. Midl. Nat.* 73: 299-35.

Johnson, D. R. 1967. Diet and reproduction of Colorado pikas. *J. Mamm.* 48: 311-15.

Kilham, L. 1958. Territorial behavior in pikas. *J. Mamm.* 39: 107.

Lutton, L. M. 1975. Notes on territorial behavior and response to predators of the pika, *Ochotona princeps. J. Mamm.* 56: 231-34.

MacArthur, R. A., and Want, C. H. 1973. Physiology of thermoregulation in the pika, *Ochotona princeps. Can. J. Zool.* 51: 11-16.

Markham, O. D., and Whicker, F. W. 1973. Notes on the behavior of the pika (*Ochotona princeps*) in captivity. *Am. Midl. Nat.* 89: 192-99.

Millar, J. S. 1972. Timing of breeding of pikas in southwestern Alberta. *Can. J. Zool.* 50: 665-69.

Millar, J. S., and Zwickel, F. C. 1972. Characteristics and ecological significance of hay piles of pikas. *Mammalia* 36: 657-67.

Orr, R. T. 1977. *The little known pika.* New York: Macmillan.

Severald, J. H. 1950. The gestation period of the pika (*Ochotona princeps*). *J. Mamm.* 31: 356-57.

Somers, P. 1973. Dialects in southern Rocky Mountain pikas, *Ochotona princeps* (Lagomorpha). *Anim. Behav.* 21: 124-37.

Svendsen, G. E. 1979. Territoriality and behavior in a population of pikas (*Ochotona princeps*). *J. Mamm.* 60: 324-30.

Tyser, R. W. 1980. Use of substrate for surveillance behaviors in a community of talus slope mammals. *Am. Midl. Nat.* 104: 32-38.

## Cottontails

Anderson, R. M. 1940. The spread of cottontail rabbits in Canada. *Can. Field-Nat.* 54: 70-72.

*Casteel, D. A. 1966. Nest building, parturition, and copulation in the cottontail rabbit. *Am. Midl. Nat.* 75: 160-67.

Conaway, C. H.; Sadler, K. C.; and Hazelwood, D. H. 1974. Geographic variation in litter size and onset of breeding in cottontails. *J. Wildl. Mgmt.* 38: 473-81.

Conaway, C. H.; Wight, H. M.; and Sadler, K. C. 1963. Annual production by a cottontail population. *J. Wildl. Mgmt.* 27: 171-75.

Kirkpatrick, C. M. 1950. Crow predation upon nestling cottontails. *J. Mamm.* 31: 322-27.

McKay, D. O., and Verts, B. J. 1978. Estimates of some attributes of a population of Nuttall's cottontails. *J. Wildl. Mgmt.* 42: 159-68.

Marsden, H. M., and Conaway, C. H. 1963. Behavior and the reproductive cycle in the cottontail. *J. Wildl. Mgmt.* 27: 161-70.

*Marsden, H. M., and Holler, N. R. 1964. *Social behavior in confined populations of the cottontail and the swamp rabbit.* Wildl. Monogr. No. 13.

## Hares

Aldous, C. M. 1937. Notes on the life history of the snowshoe hare. *J. Mamm.* 18: 46-57.

Bider, J. R. 1974. Snowshoes in the north woods. *Nature Can.* 3(1): 20-23.

Brand, C. J.; Vowles, R. H.; and Keith, L. B. 1975. Snowshoe hare mortality monitored by telemetry. *J. Wildl. Mgmt.* 39: 741-47.

Brand, C. J. and Keith, L. B. 1979. Lynx demography during a snowshoe hare decline in Alberta. *J. Wildl. Mgmt.* 43: 827-49.

Criddle, S. 1938. A study of the snowshoe rabbit. *Can. Field-Nat.* 52: 31-40.

Forcum, D. L. 1966. Postpartum behavior and vocalizations of snowshoe hares. *J. Mamm.* 47: 543.

Grange, W. B. 1932. Observations on the snowshoe hare, *Lepus americanus phaeonatus* Allen. *J. Mamm.* 13: 1-19.

Houston, C. S. 1975. Reproduction in great horned owls. *Bird-Banding* 46: 302-4.

Keith, L. B. 1964. Daily activity pattern of snowshoe hares. *J. Mamm.* 45: 626-27.

———. 1974a. Some features of population dynamics in mammals. In *Proceedings of the*

*XIth international congress of game biologists*, ed. S. Lundstrom, pp. 17-58. Stockholm, Sweden: National Swedish Environment Protection Board.

———. 1974b. *Snowshoe hare.* Ottawa: Canadian Wildlife Service Hinterland Who's Who series.

Keith, L. B., and Meslow, E. C. 1966. Animals using runways in common with snowshoe hares. *J. Mamm.* 47: 541.

Keith, L. B., and Windberg, L. A. 1978. *A demographic analysis of the snowshoe hare cycle.* Wildl. Monogr. No. 58.

McInvaille, W. B., and Keith, L. B. 1974. Predator prey relations and breeding biology of the great horned owl and red-tailed hawk in Central Alberta. *Can. Field-Nat.* 88: 1-20.

Meslow, E. C., and Keith, L. B. 1968. Demographic parameters of a snowshoe hare population. *J. Wildl. Mgmt.* 32: 812-34.

Pruitt, W. O., Jr. 1975. Life in the snow. *Nature Can.* 4(4): 40-44.

Rongstad, O. J., and Tester, J. R. 1971. Behavior and maternal relations of young snowshoe hares. *J. Wildl. Mgmt.* 35: 338-46.

Walkinshaw, L. W. 1947. Notes on the arctic hare. *J. Mamm.* 28: 353-57.

Wolff, J. O. 1980. The role of habitat patchiness in the population dynamics of snowshoe hares. *Ecol. Monogr.* 50: 111-30.

## Introduction to rodents

Hanney, P. W. 1975. *Rodents: their lives and habits.* London: David and Charles.

Wood, A. E. 1974. Rodentia. *Encyclopaedia Britannica Macropaedia* 15: 969-80.

## Least Chipmunk

Brand, L. R. 1976. the vocal repertoire of chipmunks (genus *Eutamias*) in California. *Anim. Behav.* 24: 319-35.

Broodbooks, H. E. 1974. Tree nests of chipmunks with comments on associated behavior and ecology. *J. Mamm.* 55: 630-39.

*Criddle, S. 1943. The little northern chipmunk in southern Manitoba. *Can. Field-Nat.* 57: 81-86.

Gordon, K. 1943. The natural history and behavior of the western chipmunk and mantled ground squirrel. *Oregon State Monogr., Studies Zool.* 5: 1-104.

Heller, H. C., and Gates, D. M. 1971. Altitudinal zonation of chipmunks (*Eutamias*): energy budgets. *Ecol.* 52: 424-33.

Orr, L. W. 1930. An unusual chipmunk nest. *J. Mamm.* 11: 315.

Pivorun, E. B. 1976. A biotelemetry study of the thermoregulatory patterns of *Tamias striatus* and *Eutamias minimus* during hibernation. *Comp. Biochem. Physiol.* 53A: 265-71.

Shaw, W. T. 1944. Brood nests and young of two western chipmunks in the Olympic mountains of Washington. *J. Mamm.* 25: 274-84.

Sheppard, D. H. 1968a. A comparison of reproduction in two chipmunk species (*Eutamias*). *Can. J. Zool.* 47: 603-8.

———. 1968b. Seasonal changes in body and adrenal weights of chipmunks (*Eutamias*). *J. Mamm.* 49: 463-74.

———. 1971. Competition between two chipmunk species (*Eutamias*). *Ecol.* 52: 320-29.

———. 1973. *Chipmunk.* Ottawa: Canadian Wildlife Service Hinterland Who's Who series.

## Woodchuck

Bailey, E. D., and Davis, D. E. 1965. The utilization of body fat during hibernation in woodchucks. *Can. J. Zool.* 43: 701-7.

Barash, D. P. 1974. Mother-infant relations in captive woodchucks (*Marmota monax*). *Anim. Behav.* 22: 446-48.

Bronson, F. H. 1962. Daily and seasonal activity patterns in woodchucks. *J. Mamm.* 43: 425-27.

———. 1964. Agonistic behaviour in woodchucks. *Anim. Behav.* 12: 470-78.

*de Vos, A., and Gillespie, D. I. 1960. A study of woodchucks on an Ontario farm. *Can. Field-Nat.* 74: 130-45.

*Grizzel, R. A., Jr. 1955. A study of the southern woodchuck, *Marmota monax monax*. *Am. Midl. Nat.* 53: 257-93.

Hamilton, W. J., Jr. 1934. The life history of the rufescent woodchuck, *Marmota monax rufescens* Howell. *Ann. Carneg. Mus.* 23: 87-118.

Johnson, A. M. 1926. Tree climbing woodchucks again. *J. Mamm.* 7: 132-33.

Kirkpatrick, R. D.; Martin, D. N.; and McCall, J. D. 1969. Red fox and woodchuck in same burrow. *Ind. Aud.* 47: 124-25.

Lloyd, J. E. 1972. Vocalization in *Marmota monax*. *J. Mamm.* 53: 214-16.

MacClintock, D. 1970. *Squirrels of North America.* New York: Van Nostrand Reinhold.

Merriam, H. G. 1963. An unusual fox: woodchuck relationship. *J. Mamm.* 44: 115-16.

———. 1971. Woodchuck burrow distribution and related movement patterns. *J. Mamm.* 52: 732-46.

Robb, W. H. 1926. Another tree-climbing woodchuck. *J. Mamm.* 7: 133.

Schoonmaker, W. J. 1966. *The world of the woodchuck.* Philadelphia: J. B. Lippincott.

Snyder, R. L., and Christian, J. J. 1960. Reproductive cycle and litter size of the woodchuck. *Ecol.* 4: 647-56.

## Marmots

Balph, D. F., and Stokes, A. W. 1963. On the ethology of a population of Uinta ground squirrels. *Am. Midl. Nat.* 69: 106-126.

Barash, D. P. 1973. Territorial and foraging behavior of pika (*Ochotona princeps*) in Montana. *Am. Midl. Nat.* 89: 202-7.

*———. 1974. The social behaviour of the hoary marmot (*Marmota caligata*). *Anim. Behav.* 22: 256-61.

———. 1974. The evolution of marmot societies: a general theory. *Science* 185: 415-20.

———. 1975. Ecology of paternal behaviour in the hoary marmot (*Marmota caligata*): an evolutionary interpretation. *J. Mamm.* 56: 613-18.

———. 1976. Pre-hibernation behavior of free-living hoary marmots, *Marmota caligata*. *J. Mamm.* 57: 182-85.

Gray, D. R. 1975. The marmots of Spotted Nellie Ridge. *Nature Can.* 4(1): 3-8.

Munro, W. T. 1978. Status report on Vancouver Island marmot *Marmota vancouverensis* in Canada. Committee on the status of endangered wildlife in Canada.

Steiner, A. L. 1975. "Greeting" behavior in some sciuridae, from an ontogenetic, evolutionary and socio-behavioral perspective. *Nat. Can.* 102: 733-51.

Taulman, J. F. 1977. Vocalizations of the hoary marmot, *Marmota caligata*. *J. Mamm.* 58: 681-83.

Tyser, R. W. 1980. Use of substrate for surveillance behaviors in a community of talus slope mammals. *Am. Midl. Nat.* 104: 32-38.

## Ground squirrels

Betts, B. J. 1976. Behaviour in a population of Columbia ground squirrels, *Spermophilus columbianus columbianus*. *Anim. Behav.* 24: 652-80.

Carl, E. A. 1971. Population control in arctic ground squirrels. *Ecol.* 52: 395-413.

Clark, T. W. 1970. Early growth, development, and behavior of the Richardson ground squirrel (*Spermophilus richardsoni* [sic] *elegans*). *Am. Midl. Nat.* 83: 197-205.

Hisaw, F. L., and Emergy, F. E. 1927. Food selection of ground squirrels, *Citellus tridecemlineatus*. *J. Mamm.* 8: 41-44.

Johnson, A. M. 1922. An observation of the carnivorous propensities of the gray gopher. *J. Mamm.* 3: 187.

Leach, D. 1978. The prairie excavator. *Nature Can.* 7(3): 4-8.

McKeever, S. 1964. The biology of the golden-mantled ground squirrel, *Citellus lateralis*. *Ecol. Monogr.* 34: 383-401.

Melchior, H. R. 1971. Characteristics of Arctic ground squirrel alarm calls. *Oecologia* 7: 184-90.

Michener, D. R. 1972. Notes on home range and social behavior in adult Richardson's ground squirrels (*Spermophilus richardsonii*). *Can. Field-Nat.* 86: 77-79.

——— 1974. Annual cycle of activity and weight changes in Richardson's ground squirrel, *Spermophilus richardsonii*. *Can. Field-Nat.* 88: 409-13.

*Michener, G. R. 1979a. The circannual cycle of Richardson's ground squirrels in southern Alberta. *J. Mamm.* 60: 760-68.

*——— 1979b. Spatial relationships and social organization of adult Richardson's ground squirrels. *Can. J. Zool.* 57: 125-39.

——— 1979c. Yearly variations in the population dynamics of Richardson's ground squirrels. *Can. Field-Nat.* 93: 363-70.

Michener, G. R., and Michener, D. R. 1973. Spatial distribution of yearlings in a Richardson's ground squirrel population. *Ecol.* 54: 1138-42.

*——— 1977. Population structure and dispersal in Richardson's ground squirrels. *Ecol.* 58: 360-68.

Michener, G. R., and Sheppard, D. H. 1972. Social behavior between adult female Richardson's ground squirrels (*Spermophilus richardsonii*) and their own and alien young. *Can. J. Zool.* 50: 1343-49.

Quanstrom, W. R. 1971. Behaviour of Richardson's ground squirrel *Spermophilus richardsonii richardsonii*. *Anim. Behav.* 19: 646-51.

Sheppard, D. H. 1979. *Richardson's ground squirrel.* Ottawa: Canadian Wildlife Service Hinterland Who's Who series.

Sowls, L. K. 1948. The Franklin ground squirrel, *Citellus franklinii* (Sabine), and its relationship to nesting ducks. *J. Mamm.* 29: 113-37.

Steiner, A. L. 1971. Play activity of Columbian ground squirrels. *Z. Tierpsych.* 28: 247-61.

——— 1972. Mortality resulting from intraspecific fighting in some ground squirrel populations. *J. Mamm.* 53: 601-3.

Streubel, D. P., and Fitzgerald, J. P. 1978. *Spermophilus tridecemlineatus.* Mamm. Sp. No. 103.

Watton, D. G., and Keenleyside, M. H. A. 1974. Social behaviour of the arctic ground squirrel, *Spermophilus undulatus. Behav.* 50: 77-79.

Yeaton, R. I. 1972. Social behavior and social organization in Richardson's ground squirrel (*Spermophilus richardsonii*) in Saskatchewan. *J. Mamm.* 53: 139-47.

## Red Squirrel

Fancy, S. G. 1980. Nest-tree selection by red squirrels in a boreal forest. *Can. Field-Nat.* 94: 198.

Ferron, J. 1975. Solitary play of the red squirrel (*Tamiasciurus hudsonicus*). *Can. J. Zool.* 53: 1495-99.

Hardy, G. A. 1949. Squirrel cache of fungi. *Can. Field-Nat.* 63: 86-87.

Hayward, C. L. 1940. Feeding habits of the red squirrel. *J. Mamm.* 21: 220.

Kemp, G. A., and Keith, L. B. 1970. Dynamics and regulation of red squirrel (*Tamiasciurus hudsonicus*) populations. *Ecol.* 51: 763-79.

Kilham, L. 1954. Territorial behaviour of red squirrel. *J. Mamm.* 35: 252-53.

Klugh, A. B. 1927. Ecology of the red squirrel. *J. Mamm.* 8: 1-32.

Lang, H. 1925. How squirrels and other rodents carry their young. *J. Mamm.* 6: 18-24.

Millar, J. S. 1970. The breeding season and reproductive cycle of the western red squirrel. *Can. J. Zool.* 48: 471-73.

——— 1970. Variations in fecundity of the red squirrel, *Tamiasciurus hudsonicus* (Erxleben). *Can. J. Zool.* 48: 1055-58.

Northcotte, T. H. 1977. *Marten.* Ottawa: Canadian Wildlife Service Hinterland Who's Who series.

Pauls, R. W. 1978. Behavioural strategies relevant to the energy economy of the red squirrel (*Tamiasciurus hudsonicus*). *Can. J. Zool.* 56: 1519-25.

Petrides, G. S. 1941. Snow burrows of the red squirrel (*Tamiasciurus*). *J. Mamm.* 22: 393-94.

Pruitt, W. O. Jr. 1960. *Animals of the north.* New York: Harper and Row.

Pruitt, W. O. Jr., and Lucier, C. V. 1958. Winter activity of red squirrels in interior Alaska. *J. Mamm.* 39: 443-44.

Rusch, D. A., and Reeder, W. G. 1978. Population ecology of Alberta red squirrels. *Ecol.* 59: 400-420.

Smith, C. C. 1968. The adaptive nature of social organization in the genus of three [*sic*] squirrels *Tamiasciurus. Ecol. Monogr.* 38: 31-63.

——— 1970. The coevolution of pine squirrels (*Tamiasciurus*) and conifers. *Ecol. Monogr.* 40: 349-71.

Smith, M. C. 1968. Red squirrel responses to spruce cone failure in interior Alaska. *J. Wildl. Mgmt.* 32: 305-17.

Zirul, D. L., and Fuller, W. A. 1971. Winter fluctuations in size of home range of the red squirrel (*Tamiasciurus hudsonicus*). *Trans. N. Amer. Wildl. Conf.* 35: 115-27.

## Northern flying squirrel

Booth, E. S. 1946. Notes on the life history of the flying squirrel. *J. Mamm.* 27: 28-30.

Cowan, I. McT. 1936. Nesting habits of the flying squirrel, *Glaucomys sabrinus. J. Mamm.* 17: 58-60.

Coventry, A. F. 1932. Notes on the Mearns flying squirrel. *Can. Field-Nat.* 46: 75-78.

Davis, W. 1963. Reproductive ecology of the northern flying squirrel. Master's thesis, University of Saskatchewan.

Hampson, C. 1971. *Into the woods beyond.* Toronto: Macmillan.

McKeever, S. 1960. Food of the northern flying squirrel in north-eastern California. *J. Mamm.* 41: 270-71.

*Muul, I., and Alley, J. W. 1963. Night gliders of the woodlands. *Nat. Hist.* 72 (5): 18-25.

Seton, E. T. 1909. *Life-histories of northern animals: an account of the mammals of Manitoba.* New York: Charles Scribner's Sons.

*Weigl, P. D., and Osgood, D. W. 1974. Study of the northern flying squirrel, *Glaucomys sabrinus,* by temperature telemetry. *Am. Midl. Nat.* 92: 482-86.

## Beaver

Aleksiuk, M. 1968. Scent-mound communication, territoriality, and population regulation in beaver (*Castor canadensis* Kuhl). *J. Mamm.* 49: 759-62.

——— 1970a. The function of the tail as a fat storage depot in the beaver (*Castor canadensis*). *J. Mamm.* 51: 145-48.

——— 1970b. The seasonal food regime of arctic beavers. *Ecol.* 51: 264-70.

Aleksiuk, M., and Cowan, I. McT. 1969. The winter metabolic depression in Arctic beavers (*Castor canadensis* Kuhl) with comparisons to California beavers. *Can. J. Zool.* 47: 965-79.

Berry, S. S. 1923. Observations on a Montana beaver canal. *J. Mamm.* 4: 92-103.

Bradt, G. W. 1938. A study of beaver colonies in Michigan. *J. Mamm.* 19: 139-62.

Green, H. U. 1936. The beaver of the Riding Mountain, Manitoba. *Can. Field-Nat.* 50: 1-8, 21-23, 36-50, 61-67, 85-92.

Guenther, S. E. 1948. Young beavers. *J. Mamm.* 29: 419-20.

Hatt, R. T. 1944. A large beaver-felled tree. *J. Mamm.* 25: 313.

Hitchcock, H. B. 1954. Felled tree kills beaver (*Castor canadensis*). *J. Mamm.* 35: 452.

Hodgdon, H. E., and Larson, J. S. 1973. Some sexual differences in behavior within a colony of marked beavers (*Castor canadensis*). *Anim. Behav.* 21: 147-52.

Irving, L., and Orr, M. D. 1935. The diving habits of the beaver. *Science.* 82: 569.

Jenkins, S. H., and Busher, P. E. 1979. *Castor canadensis.* Mamm. Sp. No. 120.

Martin, H. T. 1892. *Castorologia.* Montreal: Wm. Drysdale.

Novakowski, N. S. 1967. The winter bioenergetics of a beaver population in northern latitudes. *Can. J. Zool.* 45: 1107-17.

Oertli, E. F. 1976. The beavers of Kananaskis Forest. *Nature Can.* 5(1): 3-8.

Northcott, T. H. 1971. Feeding habits of beaver in Newfoundland. *Oikos* 22:407-10.

Schorger, A. W. 1953. Large Wisconsin beaver. *J. Mamm.* 34: 260-61.

Shadle, A. R. 1956. The American beaver. *Anim. King.* 59: 98-104, 152-57, 181-85.

Slough, B. G. 1978. Beaver food cache structure and utilization. *J. Wildl. Mgmt.* 42: 644-46.

Stephenson, A. B. 1969. Temperatures within a beaver lodge in winter. *J. Mamm.* 50: 134-36.

Svendsen, G. E. 1980. Population parameters and colony composition of beaver (*Castor canadensis*) in Southeast Ohio. *Am. Midl. Nat.* 104: 47-56.

Tevis, L., Jr. 1950. Summer behavior of a family of beavers in New York State. *J. Mamm.* 31: 40-65.

Townsend, J. E. 1953. Beaver ecology in western Montana with special reference to movements. *J. Mamm.* 34: 459-79.

Wallace, A. F. C., and Lathbury, V. L. 1968. Culture and the beaver. *Nat. Hist.* 77(9): 58-65.

Wilsson, L. 1968. *My beaver colony.* New York: Doubleday.

## Deer mouse

*Criddle, S. 1950. The *Peromyscus maniculatus bairdii* complex in Manitoba. *Can. Field-Nat.* 64: 169-77.

Fairbairn, D. J. 1978. Dispersal of deer mice *Peromyscus maniculatus:* proximal causes and effects on fitness. *Oecologia* 32: 171-93.

Flake, L. D. 1973. Food habits of four species of rodents on a shortgrass prairie. *J. Mamm.* 54: 636-47.

Foster, D. D. 1959. Differences in behavior and temperament between two races of the deer mouse. *J. Mamm.* 40: 496-513.

Gashwiler, J. S. 1979. Deer mouse reproduction and its relationship to the tree seed crop. *Am. Midl. Nat.* 102: 95-104.

Healey, M. C. 1967. Aggression and self-regulation of population size in deermice. *Ecol.* 48: 377-92.

Horner, B. E. 1947. Paternal care of young mice of the genus *Peromyscus. J. Mamm.* 28: 31-37.

*Howard, W. E. 1949. Dispersal, amount of inbreeding, and longevity in a local population of prairie deermice on the George Reserve, southern Michigan. *Contr. Lab. Vert. Biol. U. Mich.* 43: 1-50.

King, J. A., ed. 1968. *Biology of* Peromyscus (*Rodentia*). American Society of Mammalogists, Sp. Pub. No. 2.

Metzgar, L. H. 1979. Dispersion patterns in a *Peromyscus* population. *J. Mamm.* 60: 129-45.

Mihok, S. 1979. Behavioral structure and demography of subarctic *Clethrionomys gapperi* and *Peromyscus maniculatus. Can. J. Zool.* 57: 1520-35.

Petticrew, B. G., and Sadleir, R. M. F. S. 1974. The ecology of the deer mouse *Peromyscus maniculatus* in a coastal coniferous forest. I. Population dynamics. *Can. J. Zool.* 52: 107-18.

Sadleir, R. M. F. S. 1965. The relationship between agonistic behavior and population changes in the deermouse, *Peromyscus maniculatus* (Wagner). *J. Anim. Ecol.* 34: 331-52.

——— 1974. The ecology of the deer mouse *Peromyscus maniculatus* in a coastal coniferous forest. II. Reproduction. *Can. J. Zool.* 52: 119-31.

Williams, O. 1959. Food habits of the deer mouse. *J. Mamm.* 40: 415-19.

## Bushy-tailed wood rat

Brown, J. H. 1968. Adaptation to environmental temperature in two species of woodrats, *Neotoma cinerea* and *N. albigula.* Misc. Pub. Mus. Zoo. U. Mich. No. 135.

Egoscue, H. J. 1962. The bushy-tailed wood rat: a laboratory colony. *J. Mamm.* 43: 328-37.

Escherich, P. C. 1975. Social structure in the bushy-tailed woodrat, *Neotoma cinerea. Am. Zool.* 15: 821.

——— 1977-78. Social biology of the bushy-tailed woodrat, *Neotoma cinerea. Diss. Abs. Int.* B 38: 3584.

Finley, R. B. 1958. *The wood rats of Colorado: distribution and ecology.* U. Kansas Pub. Mus. Nat. Hist. 10: 213-552.

Hamilton, W. J. Jr. 1953. Reproduction and young of the Florida woodrat, *Neotoma f. floridana. J. Mamm.* 34: 180-89.

Horváth, O. 1966. Observation of parturition and maternal care of the bushy-tailed wood rat (*Neotoma cinerea occidentalis* Baird). *Murrelet* 47: 6-8.

## Meadow vole

Bailey, V. 1924. Breeding, feeding, and other life habits of meadow mice (*Microtus*). *J. Agr. Res.* 27: 523-35.

Fish, P. G. 1974. Notes on the feeding habits of *Microtus ochrogaster* and *M. pennsylvanicus. Am. Midl. Nat.* 92: 460-61.

Getz, L. L. 1961. Home ranges, territoriality, and movement of the meadow vole. *J. Mamm.* 42: 24-36.

——— 1962. Aggressive behavior of the meadow and prairie voles. *J. Mamm.* 43: 351-58.

Golley, F. B. 1961. Interaction of natality, mortality and movement during one annual cycle in a *Microtus* population. *Am. Midl. Nat.* 66: 152-59.

Hamilton, W. J., Jr. 1940. Life and habits of field mice. *Sci. Monthly* 50: 425-34.

Iverson, S. L., and Turner, B. N. 1972. Winter coexistence of *Clethrionomys gapperi* and *Microtus pennsylvanicus* in a grassland habitat. *Am. Midl. Nat.* 88: 440-45.

Keller, B. L., and Krebs, C. J. 1970. *Microtus* population biology; III. Reproductive changes in fluctuating populations of *M. ochrogaster* and *M. pennsylvanicus* in southern Indiana, 1965-67. *Ecol. Monogr.* 40: 263-94.

Krebs, C. J. 1970. *Microtus* population biology: behavioral changes associated with the population cycle in *M. ochrogaster* and *M. pennsylvanicus. Ecol.* 51: 34-52.

Krebs, C. J.; Keller, B. L.; and Tamarin, R. H. 1969. Microtus population biology; demographic changes in fluctuating populations of *M. ochrogaster* and *M. pennsylvanicus* in southern Indiana. *Ecol.* 50: 587-607.

Krebs, C. J., and Myers, J. H. 1974. Population cycles in small mammals. In *Advances in ecological research,* vol. 8, ed. A. Macfadyen, pp. 267-399. New York: Academic Press.

Kucera, E., and Fuller, W. A. 1978. A winter study of small rodents in aspen parkland. *J. Mamm.* 59: 200-204.

Mallory, F. F., and Clulow, F. V. 1977. Evidence of pregnancy failure in the wild meadow vole, *Microtus pennsylvanicus. Can. J. Zool.* 55: 1-17.

Morris, R. D. 1969. Competitive exclusion between *Microtus* and *Clethrionomys* in the aspen parkland of Saskatchewan. *J. Mamm.* 50: 291-301.

Rose, R. K. 1979. Levels of wounding in the meadow mole, *Microtus pennsylvanicus. J. Mamm.* 60: 37-45.

Tamarin, R. H. 1977. Dispersal in island and mainland voles. *Ecol.* 58: 1044-54.

Thompson, D. Q. 1965. Food preferences of the meadow vole (*Microtus pennsylvanicus*) in relation to habitat affinities. *Am. Midl. Nat.* 74: 76-86.

To, L. P., and Tamarin, R. H. 1977. The relation of population density and adrenal gland weight in cycling and noncycling voles (*Microtus*). *Ecol.* 58: 928-34.

Turner, B. N., and Iverson, S. L. 1973. The annual cycle of aggression in male *Microtus pennsylvanicus,* and its relation to population parameters. *Ecol.* 54: 967-81.

Turner, B. N.; Perrin, M. R.; and Iverson, S. L. 1974. Winter coexistence of voles in spruce forest: relevance of seasonal changes in aggression. *Can. J. Zool.* 53: 1004-11.

Weilert, N. G., and Shump, K. A., Jr. 1976. Physical parameters of *Microtus* nest construction. *Trans. Kans. Acad. Sci.* 79: 161-64.

## Muskrat

Aleksiuk, M. 1974. *Muskrat.* Ottawa: Canadian Wildlife Service Hinterland Who's Who series.

Bellrose, F. C. 1950. The relationship of muskrat populations to various marsh and aquatic plants. *J. Wildl. Mgmt.* 14: 299-315.

Dozier, H. L. 1948. Estimating muskrat populations by house count. *Trans. N. Am. Wildl. Conf.* 13: 372-89.

Errington, P. L. 1961. *Muskrats and marsh management.* Harrisburg, Penn.: Stackpole.

*——— 1963. *Muskrat populations.* Ames, Iowa: Iowa State University Press.

Fuller, W. A. 1951. *Natural history and economic*

*importance of the muskrat in the Athabasca-Peace Delta, Wood Buffalo Park.* Ottawa: Wildl. Mgmt. Bull. Ser. 1 No. 2.

Johansen, K. 1962. Buoyancy and insulation in the muskrat. *J. Mamm.* 43: 64-68.

Le Boulenge, E. 1972. *Etat de nos connaissances sur l'écologie du rat musque Ondatra zibethica L. Terre Vie* 26: 3-37.

Macarthur, R. A. 1979. Winter movements and home range of the muskrat. *Can. Field-Nat.* 92: 345-49.

*Macarthur, R. A., and Aleksiuk, M. 1979. Seasonal microenvironments of the muskrat (*Ondatra zibethicus*) in a northern marsh. *J. Mamm.* 60: 146-54.

Sprugel, G., Jr. 1954. Spring dispersal and settling activities of central Iowa muskrats. *Iowa State Univ. J. Sci.* 26: 71-84.

Stevens, W. E. 1953. *The northwestern muskrat of the Mackenzie Delta, Northwest Territories, 1947-48.* Ottawa: Wildl. Mgmt. Bull., Ser. 1 No. 8.

Willner, G. R.; Feldhamer, G. A.; Zucker, E. E.; and Chapman, J. A. 1980. *Ondatra zibethicus.* Mamm. Sp. No. 141.

## Porcupine

Batchelder, C. F. 1948. Notes on the Canada porcupine. *J. Mamm.* 29: 260-68.

Curtis, J. D., and Kozkicky, E. L. 1944. Observations on the eastern porcupine. *J. Mamm.* 25: 137-46.

Gabrielson, I. N. 1928. Notes on the habits and behavior of the porcupine in Oregon. *J. Mamm.* 9: 33-38.

Marshall, H. 1951. Accidental death of a porcupine. *J. Mamm.* 32: 221.

Murie, O. J. 1926. The porcupine in northern Alaska. *J. Mamm.* 7: 109-13.

Quick, H. F. 1953. Occurrence of porcupine quills in carnivorous mammals. *J. Mamm.* 34: 257-59.

Saunders, A. A. 1932. The voice of the porcupine. *J. Mamm.* 13: 167-68.

Schoonmaker, W. J. 1938. The fisher as a foe of the porcupine in New York State. *J. Mamm.* 19: 373-74.

Seton, E. T. 1932. The song of the porcupine (*Erethizon epixanthum*). *J. Mamm.* 13: 168-69.

Shadle, A. R. 1946. Copulation in the porcupine. *J. Wildl. Mgmt.* 10: 159-62.

——— 1947. Porcupine spine penetration. *J. Mamm.* 28: 180-81.

——— 1955. Removal of foreign quills by porcupines. *J. Mamm.* 36: 463-65.

Shadle, A. R., and Ploss, W. R. 1943. An unusual porcupine parturition and development of the young. *J. Mamm.* 24: 492-96.

Shadle, A. R.; Smelzer, M.; and Metz., M. 1946. The sex reactions of porcupines (*Erethizon d. dorsatum*) before and after copulation. *J. Mamm.* 27: 116-21.

Shapiro, J. 1949. Ecological and life history notes on the porcupine in the Adirondacks. *J. Mamm.* 30: 247-57.

Wade, O. 1931. The voice of the porcupine. *J. Mamm.* 12: 71.

Woods, C. A. 1973. *Erethizon dorsatum.* Mamm. Sp. No. 29.

## Introduction to whales, dolphins, and porpoises

Beamish, P. 1978. Evidence that a captive humpback whale (*Megaptera novaeangliae*) does not use sonar. *Deep-Sea Res.* 25: 469-72.

Haley, D., ed. 1978. *Marine mammals of eastern North Pacific and Arctic waters.* Seattle: Pacific Search Press.

Norris, K. S. 1974. Whale. *Encyclopaedia Britannica Macropaedia* 19: 805-10.

## Killer whale

Baldridge, A. 1972. Killer whales attack and eat a gray whale. *J. Mamm.* 53: 898-900.

Barr, N., and Barr, L. 1972. An observation of killer whale predation on a Dall porpoise. *Can. Field-Nat.* 86: 170-71.

Brown, D. H., and Norris, K. S. 1956. Observations of captive and wild cetaceans. *J. Mamm.* 37: 311-26.

Caldwell, M. C., and Caldwell, D. K. 1966. Epimeletic (care-giving) behavior in Cetacea. In *Whales, dolphins, and porpoises,* ed. K. S. Norris, pp. 755-89. Berkeley: University of California Press.

*Chandler, R.; Goebel, C.; and Balcomb, K. 1977. Who is that killer whale: a new key to whale watching. *Pac. Search* 11: 25-28.

Denniston, G. C. 1973. Killer whale behavior with young. *Murrelet* 54(2): 22.

Hancock, D. 1965. Killer whales kill and eat a minke whale. *J. Mamm.* 46: 341-42.

Hoyt, E. 1977. *Orcinus orca:* separating fact from fantasies. *Oceans* 10(4): 23-26.

Jacobsen, J. 1980. A synopsis: the wild birth of a killer whale (*Orcinus orca*). 3 pp. Mimeographed.

Jonsgard, A. 1968. A note on the attacking behaviour of the killer whale (*Orcinus orca*). *Nor. Hval.-Tid.* 57: 84-85.

Martinez, D. R., and Klinghammer, E. 1970. The behavior of the whale *Orcinus orca:* a review of the literature. *Z. Tierpsych.* 27: 828-39.

Newman, M. A., and McGeer, P. L. 1966. The capture and care of a killer whale, *Orcinus orca,* in British Columbia. *Zoologica* 51: 59-69.

Notarbartolo di Sciara, G. 1977. A killer whale (*Orcinus orca* L.) attacks and sinks a sailing boat. *Nature* 68: 218-20.

Pike, G. C., and MacAskie, I. B. 1969. *Marine mammals of British Columbia.* Fish. Res. Bd. Can. Bull. 171.

Rice, D. W. 1968. Stomach contents and feeding behavior of killer whales in the eastern north Pacific. *Nor. Hval.-Tid.* 1: 35-38.

Salisbury, D. F. 1978. A whale called killer. *Nat. Wildl.* 16(2): 4-9.

Scheffer, V. B. 1971. Fat-choppers — killer whales. In *Toothed whales in eastern North Pacific and Arctic waters,* ed. A. Seed, pp. 13-18. Seattle: Pacific Search.

*———. 1978. Killer whale. In *Marine mammals of eastern North Pacific and Arctic waters,* ed. D. Haley, pp. 120-27. Seattle: Pacific Search Press.

Scheffer, V. B., and Slipp, J. W. 1948. The whales and dolphins of Washington state with a key to the cetaceans of the west coast of North America. *Am. Midl. Nat.* 39: 257-337.

Schevill, W. E., and Watkins, W. A. 1966. Sound structure and directionality in *Orcinus* (killer whale). *Zoologica* 51: 71-76.

Spong, P.; Bradford, J.; and White, D. 1970. Field studies of the behavior of the killer whale *(Orcinus orca). Proc. 7th Ann. Conf. Biol. Sonar Diving Mamm.* 7: 169-83.

Steiner, W. W.; Hain, J. H.; Winn, H. E.; and Perkins, P. J. 1979. Vocalizations and feeding behavior of the killer whale (*Orcinus orca*). *J. Mamm.* 60: 823-27.

Tarpy, C. 1979. Killer whale attack! *Nat. Geog.* 155: 542-45.


## Porpoises


Andersen, S., and Dziedzic, A. 1964. Behavior patterns of captive harbor porpoise *Phocaena phocaena* L. *Bull. Inst. Oceanogr. Monaco* 63: 1-20.

Caldwell, D. K., and Caldwell, M. C. 1971. From romance to research — dolphins and porpoises. In *Toothed whales in eastern North Pacific and Arctic waters,* ed. A. Seed, pp. 4-12. Seattle: Pacific Search.

Cowan, I. McT. 1944. The Dall porpoise, *Phocoenoides dalli* (True), of the northern Pacific Ocean. *J. Mamm.* 25: 295-306.

Dudok van Heel, W. H. 1959. Audio-direction finding in the porpoise *(Phocoena phocoena). Nature* 183: 1063.

Evans, W. E. 1973. Echolocation by marine delphinids and one species of fresh-water dolphin. *J. Acous. Soc. Am.* 54: 191-99.

Fink, B. D. 1959. Observation of porpoise predation on a school of Pacific sardines. *Calif. Fish Game* 45: 216-17.

Gaskin, D. E.; Arnold, P. W.; and Blair, B. A. 1974. *Phocoena phocoena.* Mamm. Sp. No. 42.

*Leatherwood, S., and Reeves, R. R. 1978. Porpoises and dolphins. In *Marine mammals of eastern North Pacific and Arctic waters,* ed. D. Haley, pp. 96-111. Seattle: Pacific Search Press.

Linehan, E. J. 1979. The trouble with dolphins. *Nat. Geog.* 155: 506-41.

*Morejohn, G. V. 1979. The natural history of Dall's porpoise in the North Pacific Ocean. In *Behavior of marine animals: current perspectives in research,* vol. 3, eds. H. E. Winn and B. L. Olla, pp. 45-83. New York: Plenum Press.

Norris, K. S., ed. 1966. *Whales, dolphins, and porpoises.* Berkeley: University of California Press.

Norris, K. S. 1974. Whale. *Encyclopaedia Britannica Macropaedia.* 19: 805-10.

Orr, R. T. 1937. A porpoise chokes on a shark. *J. Mamm.* 18: 370.

Pike, G. C., and MacAskie, I. B. 1969. *Marine mammals of British Columbia.* Fish. Res. Bd. Can. Bull. 171.

Rae, B. B. 1965. The food of the common porpoise (*Phocaena phocaena*). *J. Zool.* 146: 114-22.

Scheffer, V. B. 1949. The Dall porpoise, *Phocoenoides dalli,* in Alaska. *J. Mamm.* 30: 116-21.

*Scheffer, V. B., and Slipp, J. W. 1948. The whales and dolphins of Washington state with a key to the cetaceans of the west coast of North America. *Am. Midl. Nat.* 39: 257-337.

Schevill, W. E.; Watkins, W. A.; and Ray, C. 1969. Click structure in the porpoise, *Phocoena phocoena. J. Mamm.* 50: 721-28.

Sergeant, D. E. 1969. Feeding rates of Cetacea. *FiskDir. Skar. Ser. HavUnders* 15: 246-58.


## Gray whale


Baldridge, A. 1972. Killer whales attack and eat a gray whale. *J. Mamm.* 53: 898-900.

Darling, J. 1977. The Vancouver Island gray whales. *Waters* (J. Vancouver Aquar.) 2(1): 5-19.

Hart, F. G. 1977. Observations on the spring migration and behavior of gray whales near Pachena Point, British Columbia. *Murrelet* 58(2): 40-43.

Killingley, J. S. 1979. Migrations of California gray whales tracked by oxygen — 18 variations in their epizoic barnacles. *Science* 207: 759-60.

Morejohn, G. C. 1968. A killer whale–gray whale encounter. *J. Mamm.* 49: 327-28.

Pike, G. C. 1962. Migration and feeding of the gray whale (*Eschrichtius gibbosus*). *J. Fish. Res. Bd. Canada* 19: 815-38.

*Rice, D. W. 1978. Gray whale. In *Marine mammals of eastern North Pacific and Arctic waters,* ed. D. Haley, pp. 54-61. Seattle: Pacific Search Press.

Rice, D. W., and Wolman, A. A. 1971. *The life history and ecology of the gray whale* (Eschrichtius robustus). American Society of Mammalogists Special Publication No. 3.

Samaras, W. F. 1974. Reproductive behavior of the gray whale *Eschrichtius robustus,* in Baja, California. *Bull. S. Calif. Acad. Sci.* 73: 57-64.

Sund, P. N. 1975. Evidence of feeding during migration and of an early birth of the California gray whale (*Eschrichtius robustus*). *J. Mamm.* 56: 265-66.

## Humpback whale

Caldwell, M. C., and Caldwell, D. K. 1966. Epimeletic (care-giving) behavior in Cetacea. In *Whales, dolphins, and porpoises,* ed. K. S. Norris, pp. 755-89. Berkeley: University of California Press.

Chittleborough, R. G. 1958. The breeding cycle of the female humpback whale *Megaptera nodosa* (Bonnaterre). *Aust. J. Mar. Freshwater Res.* 9: 1-18.

Earle, S. A. 1979. The gentle whales. *Nat. Geog.* 155: 2-17.

Edel, R. K., and Winn, H. E. 1978. Observations on underwater locomotion and flipper movement of the humpback whale *Megaptera novaeangliae. Mar. Biol.* 48: 279-87.

Haley, D., ed. 1978. *Marine Mammals of eastern North Pacific and Arctic waters.* Seattle: Pacific Search Press.

Lockley, R. M. 1979. *Whales, dolphins, and porpoises.* London: David and Charles.

Lockyer, C. 1979. Body weights of some species of large whale. *J. Cons. Int. Explor. Mer.* 36: 259-73.

Norris, K. S. 1969. The echolocation of marine mammals. In *The Biology of marine mammals,* ed. H. T. Anderson, pp. 391-423. New York: Academic Press.

Payne, R. 1979. Humpbacks: their mysterious songs. *Nat. Geog.* 155: 18-25.

Payne, R. S., and McVay, S. 1971. Songs of humpback whales. *Science* 173: 585-97.

Pike, G. C. 1953. Colour pattern of humpback whales from the coast of British Columbia. *J. Fish. Res. Bd. Can.* 10: 320-25.

Pike, G. C., and MacAskie, I. B. 1969. *Marine mammals of British Columbia.* Fish. Res. Bd. Can. Bull. 171.

Scheffer, V. B. 1976. Exploring the lives of whales. *Nat. Geog.* 150: 752-67.

Thompson, T. J.; Winn, H. E.; and Perkins, P. J. 1979. Mysticete sounds. In *Behavior of marine mammals: current perspectives in research,* vol. 3, eds. H. E. Winn and B. L. Olla, pp. 403-431. New York: Plenum Press.

Watkins, W. A., and Schevill, W. E. 1979. Aerial observation of feeding behavior in four baleen whales: *Eubalaena glacialis, Balaenoptera borealis, Megaptera novaeangliae,* and *Balaenoptera physalus. J. Mamm.* 60: 155-63.

Winn, H. E., and Perkins, P. J. 1971. Sounds of the humpback whale. *Proc. 7th Ann. Conf. Biol. Sonar Diving Mamm.* 7: 39-52.

## Introduction to carnivores

DeBlaze, A. F., and Martin, R. E. 1974. *A manual of mammalogy, with keys to families of the world.* Dubuque, Iowa: Wm. C. Brown.

*Ewer, R. F. 1973. *The carnivores.* Ithaca, N.Y.: Cornell University Press.

Munro, W. T. 1978. Status report on sea otter *Enhydra lutris* in Canada. Committee on the Status of Endangered Wildlife in Canada.

Stains, H. J. 1974. Carnivora. *Encyclopaedia Britannica Macropaedia* 3: 926-44.

## Coyote

Alldredge, A. W., and Arthur, W. J., III. 1980. Observations on coyote–mule deer interactions at Rocky Flats, Colorado. *Am. Midl. Nat.* 103: 200-201.

Andelt, W. F.; Althoff, D. P.; and Gipson, P. S. 1979. Movements of breeding coyotes with emphasis on den site relationships. *J. Mamm.* 60: 568-75.

Bekoff, M. 1977. *Canis latrans.* Mamm. Sp. No. 79.

———, ed. 1978. *Coyotes: biology, behavior, and management.* New York: Academic Press.

Bekoff, M., and Wells, M. C. 1980. The social ecology of coyotes. *Sci. Am.* 242 (4): 130-48.

Dobie, J. F. 1950. *The voice of the coyote.* London: Hammond, Hammond.

*Gier, H. T. 1975. Ecology and behavior of the coyote (*Canis latrans).* In *The wild canids, their systematics, behavioral ecology and evolution,* ed. M. W. Fox, pp. 247-62. New York: Van Nostrand Reinhold.

Hamlin, K. L. and Schweitzer, L. L. 1979. Cooperation by coyote pairs attacking mule deer fawns. *J. Mamm.* 60: 849-50.

Lehner, P. N. 1978. Coyote vocalization: a lexicon and comparisons with other canids. *Anim. Behav.* 26: 712-22.

McMahan, P. 1975. The victorious coyote. *Nat. Hist.* 84 (1): 42-51.

———. 1978. Natural history of the coyote. In *Wolf and man: evolution in parallel,* eds. R. L. Hall and H. S. Sharp, pp. 41-54. New York: Academic Press.

Ozoga, J. J., and Harger, E. M. 1966. Winter activities and feeding habits of northern Michigan coyotes. *J. Wildl. Mgmt.* 30: 809-18.

Rathbun, A. P.; Wells, M. C.; and Bekoff, M. 1980. Cooperative predation by coyotes on badgers. *J. Mamm.* 61: 375-76.

*Ryden, H. 1975, 1979. *God's dog: a celebration of the North American coyote.* New York: Viking Press.

White, M. 1973. Description of remains of deer fawns killed by coyotes. *J. Mamm.* 54: 291-93.


## Wolf


Ballenberghe, V. V.; Erickson, A. W.; and Byman, D. 1975. *Ecology of the timber wolf in northeastern Minnesota.* Wildl. Monogr. No. 43.

Bromley, R. G. 1973. Fishing behavior of a wolf on the Taltson River, Northwest Territories. *Can. Field-Nat.* 87: 301-3.

Hall, R. L., and Sharp, H. S., eds. 1978. *Wolf and man: evolution in parallel.* New York: Academic Press.

Harrington, F. H., and Mech, L. D. 1979. Wolf howling and its role in territory maintenance. *Behav.* 68: 209-49.

*Mech, L. D. 1970. *The wolf: the ecology and behavior of an endangered species.* New York: Natural History Press.

———. 1974. *Canis lupus.* Mamm. Sp. No. 37.

———. 1977a. Productivity, mortality, and population trends of wolves in northeastern Minnesota. *J. Mamm.* 58: 560-74.

———. 1977b. Where can the wolf survive? *Nat. Geog.* 152: 518-37.

Munthe, K., and Hutchinson, J. H. 1978. A wolf-human encounter on Ellesmere Island, Canada. *J. Mamm.* 59: 876-78.

Peters, R. P., and Mech, L. D. 1975. Scent-marking in wolves. *Am. Sci.* 63: 628-37.

Peterson, R. L. 1947. A record of a timber wolf attacking a man. *J. Mamm.* 28: 294-95.

Rothman, R. J., and Mech, L. D. 1979. Scent-marking in lone wolves and newly formed pairs. *Anim. Behav.* 27: 750-60.

Theberge, J. B. 1973. Wolf management in Canada through a decade of change. *Nature Can.* 2(1): 3-10.

Weaver, J. L. 1979. Wolf predation upon elk in the Rocky Mountain parks of North America: a review. In *North American elk: ecology, behavior, and management,* eds. M. S. Boyce and L. D. Hayden-Wing, pp. 29-33. University of Wyoming.

Zimmen, E. 1975. Social dynamics of the wolf pack. In *The wild canids: their systematics, behavioral ecology and evolution,* ed. M. W. Fox, pp. 336-62. New York: Van Nostrand Reinhold.


## Arctic fox


Chesemore, D. L. 1968a. Distribution and movements of white foxes in northern and western Alaska. *Can. J. Zool.* 46: 849-54.

———. 1968b. Notes on the food habits of arctic foxes in northern Alaska. *Can. J. Zool.* 46: 1127-30.

———. 1969. Den ecology of the arctic fox in northern Alaska. *Can. J. Zool.* 47: 121-29.

Danilov, D. N. 1958. Den sites of the arctic fox (*Alopex lagopus)* in the east part of Bol'shezemel'skaya tundra. *Problems of the North.* 2: 223-29.

Eberhardt, L. E., and Hanson, W. C. 1978. Long-distance movements of arctic foxes tagged in northern Alaska. *Can. Field-Nat.* 92: 386-89.

Elton, C. 1949. Movements of arctic fox populations in the region of Baffin Bay and Smith Sound. *Polar Rec.* 5: 296-305.

Fetherston, K. 1947. Geographic variation in the incidence of occurrence of the blue phase of the arctic fox in Canada. *Can. Field-Nat.* 61: 15-18.

Garrott, R. and D. 1979. On the trail of the arctic fox. *Nat. Wildl.* 17(2): 42-47.

Krebs, C. J. 1973. *Lemmings.* Ottawa: Canadian Wildlife Service Hinterland Who's Who series.

Macinnes, C. D., and Misra, R. K. 1972. Predation on Canada goose nests at McConnell River,

Northwest Territories. *J. Wildl. Mgmt.* 36: 414-22.

Macpherson, A. H. 1969. *The dynamics of Canadian arctic fox populations.* Ottawa: Canadian Wildlife Service Report Series. No. 8.

Segal, A. N.; Popvich, T. V.; and Vain-Rib, M. A. 1976. Some of the ecological and physiological features of the arctic fox *Alopex lagopus. Zool. Zhur.* 55: 741-54.

Skrobov, V. D., and Shirokovskaya, E. A. 1967. The role of the arctic fox in improving the vegetation cover of the tundra. *Problems of the North* 11: 123-28.

Smith, T. G. 1976. Predation of ringed seal pups (*Phoca hispida*) by the arctic fox (*Alopex lagopus*). *Can. J. Zool.* 54: 1610-16.

Soper, J. D. 1944. Mammals of Baffin Island. *J. Mamm.* 25: 221-254.

Speller, S. W. 1977. *Arctic fox.* Ottawa: Canadian Wildlife Service Hinterland Who's Who series.

Stirling, I., and Archibald, W. R. 1977. Aspects of predation of seals by polar bears. *J. Fish. Res. Bd. Can.* 34: 1126-29.

Wrigley, R. E., and Hatch, D. R. M. 1976. Arctic fox migrations in Manitoba. *Arctic* 29: 147-58.

## Red fox

Ables, E. D. 1965. An exceptional fox movement. *J. Mamm.* 46: 102.

———: 1969. Activity studies of red foxes in southern Wisconsin. *J. Wildl. Mgmt.* 33: 145-53.

Butler, L. 1947. The genetics of the color phases of the red fox in the Mackenzie River locality. *Can. J. Res. Sec. D* 25: 190-215.

Cowan, I. McT. 1939. Geographic distribution of color phases of the red fox and black bear in the Pacific Northwest. *J. Mamm.* 19: 202-6.

Cross, E. C. 1941. Periodic fluctuations in numbers of the red fox in Ontario. *J. Mamm.* 40: 294-306.

Fox, M. W., ed. 1975. *The wild canids: their systematics, behavioral ecology and evolution.* New York: Van Nostrand Reinhold.

Hatfield, D. M. 1939. Winter food habits of foxes in Minnesota. *J. Mamm.* 20: 202-6.

Henry, J. D. 1976. Adaptive strategies in the behaviour of the red fox, *Vulpes vulpes* L. Ph.D. dissertation, University of Calgary.

———. 1977. The use of urine marking in the scavenging behavior of the red fox (*Vulpes vulpes*). *Behav.* 61: 82-106.

———. 1980. Fox hunting. *Nat. Hist.* 89 (1): 61-69.

MacDonald, D. W. 1976. Food caching by red foxes and some other carnivores. *Z. Tierpsych.* 42: 170-85.

Pils, C. M., and Martin, M. A. 1978. *Population dynamics, predator-prey relationships and management of the red fox in Wisconsin.* Madison, Wisconsin: Department of Natural Resources Tech. Bull. No. 105.

Preston, E. M. 1975. Home range defense in the red fox, *Vulpes vulpes* L. *J. Mamm.* 56: 645-52.

Schofield, R. D. 1958. Litter size and age ratios of Michigan red foxes. *J. Wildl. Mgmt.* 22: 313-15.

Scott, T. G., and Klimstra, W. D. 1955. *Red foxes and a declining prey population.* South Illinois Univ. Monogr. Ser. No. 1.

Sheldon, W. G. 1949. Reproductive behavior of foxes in New York State. *J. Mamm.* 30: 236-46.

Stanley, W. C. 1963. *Habits of the red fox in Northeastern Kansas.* University of Kansas Mus. Nat. Hist. Publ. No. 34.

Storm, G. L. 1965. Movements and activities of foxes as determined by radio-tracking. *J. Wildl. Mgmt.* 29: 1-13.

Tullar, B. F. Jr.; Berchielli, L. T.; and Saggese, E. P. 1976. Some implications of communal denning and pup adoption among red foxes in New York. *N.Y. Fish Game J.* 23: 92-95.

Vincent, R. E. 1958. Observations of red fox behavior. *Ecol.* 39: 755-57.

## Black bear

Aldous, S. E. 1937. A hibernating black bear with cubs. *J. Mamm.* 18: 466-68.

Boyer, R. H. 1949. Mountain coyotes kill yearling black bear in Sequoia National Park. *J. Mamm.* 30: 75.

DeWeese, L. R., and Pillmore, R. E. 1973. Bird nests in an aspen tree robbed by black bear. *Condor* 74: 488.

Erickson, A. W.; Nellor, J.; and Petrides, G. A. 1964. *The black bear in Michigan.* Michigan State Univ. Agr. Exp. Sta. Res. Bull. No. 4.

Franzreb, K. E., and Higgins, A. E. 1975. Possible bear predation on a yellow-bellied sapsucker nest. *Auk* 92: 817.

Hatler, D. F. 1972. Food habits of black bears in interior Alaska. *Can. Field-Nat.* 86: 17-31.

Henry, J. D., and Herrero, S. M. 1974. Social play in the American black bear: its similarity to canid social play and an examination of its identifying characteristics. *Am. Zool.* 14: 371-89.

*Herrero, S., ed. 1972. *Bears — their biology and management.* Morges, Switzerland: International Union for Conservation of Nature and Natural Resources.

Jonkel, C. J., and Miller, F. L. 1970. Recent records of black bears (*Ursus americanus)* on the barren grounds of Canada. *J. Mamm.* 51: 826-28.

*Jonkel, C. J., and Cowan, I.McT. 1971. *The black bear in the spruce-fir forest.* Wildl. Monogr. No. 27.

Mech, D. 1963. Biggest black bear ever caught. *Sci. Dig.* 54: 5-9.

Morse, M. A. 1937. Hibernation and breeding of the black bear. *J. Mamm.* 18: 460-65.

Murie, A. 1937. Some food habits of the black bear. *J. Mamm.* 18: 238-40.

Northcott, T. H., and Elsey, F. E. 1971. Fluctuations in black bear populations and their relationship to climate. *Can. Field-Nat.* 85: 123-28.

Pederson, R. J.; Adams, A. W.; and Williams, W. 1974. Black bear predation on an elk calf. *Murrelet* 55: 28.

Pelton, M. R.; Lentfer, J. W.; and Folk, G. E., eds. 1976. *Bears — their biology and management.* Morges, Switzerland: International Union for Conservation of Nature and Natural Resources.

Rogers, L. 1976. Effects of mast and berry crop failures on survival, growth, and reproductive success of black bears. *Trans. N.A. Wildl. Conf.* 41: 431-38.

Tietje, W. D., and Ruff, R. L. 1980. Denning behavior of black bears in boreal forest of Alberta. *J. Wildl. Mgmt.* 44: 858-70.

## Grizzly bear

Cole, G. F. 1972. Grizzly bear–elk relationships in Yellowstone National Park. *J. Wild. Mgmt.* 36: 556-61.

Craighead, F. C., Jr. 1979. *Track of the grizzly.* San Francisco: Sierra Club Books.

Craighead, F. C., Jr., and Craighead, J. J. 1972. *Grizzly bear prehibernation and denning activities as determined by radiotracking.* Wildl. Monogr. No. 32.

Craighead, J. J.; Hornocker, M. G.; and Craighead, F. C., Jr. 1969. Reproductive biology of young female grizzly bears, *J. Reprod. Fert., Suppl.* 6: 447-75.

Herrero, S. 1970. Human injury inflicted by grizzly bears. *Science* 170: 593-98.

Leopold, A. S. 1970. Weaning grizzly bears: a report on *Ursus arctos horribilis. Nat. Hist.* 79 (4): 94-101.

Macey, A. 1979. Status report on grizzly bear *Ursus arctos horribilis* in Canada. Committee on the Status of Endangered Wildlife in Canada.

Macpherson, A. H. 1965. The barren-ground grizzly. *Can. Aud.* 27 (1): 2-8.

Mundy, K. 1973. *Grizzly.* Ottawa: Canadian Wildlife Service Hinterland Who's Who series.

Mundy, K. R. D., and Flook, D. R. 1973. *Background for managing grizzly bears in the national parks of Canada.* Canadian Wildlife Service Report Series No. 22.

Pearson, A. M. 1972. The grizzly bear. *Can. Geog. J.* 64 (4): 116-23.

———. 1975. *The northern interior grizzly bear* Ursus arctos *L.* Canadian Wildlife Service Report Series No. 34.

Pelton, M. R.; Lentfer, J. W.; and Folk, G. E., eds. 1976. *Bears — their biology and management.* Morges, Switzerland: International Union for Conservation of Nature and Natural Resources.

Quimby, R. 1974. Grizzly bear. In *Mammal studies in northeastern Alaska with emphasis within the Canning River drainage.* CAGSL Biol. Rep. Ser. vol. 24.

Rausch, R. L. 1963. Geographic variation in size in North American brown bears, *Ursus arctos* L., as indicated by condylobasal length. *Can. J. Zool.* 41: 33-45.

Ruttan, R. A. 1974. Observations of grizzly bear in the northern Yukon Territory and Mackenzie River Basin, 1972. In *Studies of furbearers associated with proposed pipeline routes in the Yukon and Northwest Territories.* CAGSL Biol. Rep. Ser. vol. 9.

Storonov, D. 1972. Protocol at the annual brown bear fish feast. *Nat. Hist.* 81 (9): 66-73, 90-94.

## Polar bear

Bowes, G. W., and Jonkel, C. J. 1975. Presence and distribution of polychlorinated biphenyls (PCB) in arctic and subarctic marine food chains. *J. Fish. Res. Bd. Can.* 32: 2111-23.

Freeman, M. M. R. 1973. Polar bear predation on beluga in the Canadian Arctic. *Arctic* 26: 162-63.

Furnell, D. J., and Oolooyuk, D. 1980. Polar bear predation on ringed seals in ice-free water. *Can. Field-Nat.* 94: 88-89.

Harington, C. R. 1964. Polar bears and their present status. *Can. Aud.* 26 (1): 4-11.

———. 1968. *Denning habits of the polar bear* (Ursus maritimus *Phipps).* Ottawa: Canadian Wildlife Service Report Series No. 5.

———. 1973. *Polar bear.* Ottawa: Canadian Wildlife Service Hinterland Who's Who series.

Jonkel, C. J.; Kolenosky, G. B.; Robertson, R. J.; and Russel, R. H. 1972. Further notes on polar bear denning habits. In *Bears — their biology and management,* ed. S. Herrero, pp. 142-58. Morges, Switzerland: International Union for Conservation of Nature and Natural Resources.

Jonkel, C.; Smith, P.; Stirling, I.; and Kolenosky, G. B. 1976. *The present status of the polar bear in the James Bay and Belcher Islands area.* Ottawa: Canadian Wildlife Service Occasional Paper No. 26.

Knudsen, B. 1978. Time budgets of polar bears *(Ursus maritimus)* on North Twin Island, James Bay, during summer. *Can. J. Zool.* 56: 1627-28.

Larsen, T. 1978. *The world of the polar bear.* London: Hamlyn Publishing Group.

Russel, R. H. 1975. The food habits of polar bears of James Bay and southwest Hudson Bay in summer and autumn. *Arctic* 28: 117-39.

Smith, T. G., and Stirling, I. 1975. The breeding habitat of the ringed seal *(Phoca hispida).* The birth lair and associated structures. *Can. J. Zool.* 53: 1297-1305.

Stirling, I. 1974. Midsummer observations on the behavior of wild polar bears *(Ursus maritimus). Can. J. Zool.* 52: 1191-98.

———. 1974. Polar bear research in the Beaufort Sea. In *The coast and shelf of the Beaufort Sea,* eds. J. C. Reed and J. E. Sater, pp. 721-33. Arlington, Virginia: Arctic Institute of North America.

Stirling, I., and Jonkel, C. 1972. The great white bears. *Nature Can.* 1(3): 15-18.

Stirling, I.; Jonkel, C.; Smith, P.; Robertson, R.; and Cross, D. 1977. *The ecology of the polar bear* (Ursus maritimus) *along the western coast of Hudson Bay.* Ottawa: Canadian Wildlife Service Occasional Paper No. 33.

Stirling, I., and Latour, P. B. 1978. Comparative hunting abilities of polar bear cubs of different ages. *Can. J. Zool.* 56: 1768-72.

## Raccoon

Bergtold, W. H. 1925. Unusual nesting of a raccoon. *J. Mamm.* 6: 280-81.

Berner, A., and Gysel, L. W. 1967. Raccoon use of large tree cavities and ground burrows. *J. Wildl. Mgmt.* 31: 706-14.

Butterfield, R. T. 1954. Some raccoon and groundhog relationships. *J. Wildl. Mgmt.* 18: 433-37.

Dorney, R. S. 1954. Ecology of marsh raccoons. *J. Wildl. Mgmt.* 18: 217-25.

Ewer, R. F. 1973. *The carnivores.* Ithaca, N.Y.: Cornell University Press.

Fritzell, E. K. 1977. Dissolution of raccoon sibling bonds. *J. Mamm.* 58: 427-28.

———. 1979. Habitat use by prairie raccoons during the waterfowl breeding season. *J. Wildl. Mgmt.* 42: 118-27.

Fritzell, E. K., and Matthews, J. W. 1975. A large raccoon litter. *Prairie Nat.* 7: 87-88.

Giles, L. W. 1940. Food habits of the raccoon in eastern Iowa. *J. Wildl. Mgmt.* 4: 375-82.

Hancock, L. 1977. Here come the raccoons — full of fun! *Can. Geog. J.* 94(2): 38-45.

Houston, C. S., and Houston, M. I. 1973. A history of raccoons in Saskatchewan. *Blue Jay* 31: 103-4.

Llewellyn, L. M., and Webster, C. G. 1960. Raccoon predation on waterfowl. *Tran. N. Am. Wildl. Conf.* 25: 180-85.

Lotze, J.-H., and Anderson, S. 1979. *Procyon lotor.* Mamm. Sp. No. 119.

Lynch, G. M. 1971. Raccoons increasing in Manitoba. *J. Mamm.* 52: 621-22.

———. 1974. Some den sites of Manitoba raccoons. *Can. Field-Nat.* 88: 494-95.

Mech, L. D.; Tester, J. R.; and Warner, D. W. 1966. Fall daytime resting habits of raccoons as determined by telemetry. *J. Mamm.* 47: 450-66.

Mech, L. D., and Turkowski, F. J. 1966. Twenty-three raccoons in one winter den. *J. Mamm.* 47: 529-30.

Schneider, D. 1973. The adaptable raccoon. *Nat. Hist.* 82 (7): 64-71.

Schneider, D. G.; Mech, L. D.; and Tester, J. R. 1971. Movements of female raccoons and their young as determined by radio-tracking. *Anim. Behav. Monogr.* 4 (a): 1-43.

Schnell, J. H. 1969. Rest site selection by radio-tagged raccoons. *J. Minn. Acad. Sci.* 36: 83-88.

Sharp, W. M., and Sharp, L. H. 1956. Nocturnal movements and behavior of wild raccoons at a winter feeding station. *J. Mamm.* 37: 170-77.

Shirer, H. W., and Fitch, H. S. 1970. Comparison from radiotracking of movements and denning habits of the raccoon, striped skunk, and opossum in northeastern Kansas. *J. Mamm.* 51: 491-503.

Smith, H. C. 1972. Some recent records of Alberta mammals. *Blue Jay* 30: 53-54.

Sowls, L. K. 1949. Notes on the raccoon *(Procyon lotor hirtus)* in Manitoba. *J. Mamm.* 30: 313-14.

Stuewer, F. W. 1943. Raccoons: their habits and management in Michigan. *Ecol. Monogr.* 13: 203-57.

Sutton, R. W. 1964. Range extension of raccoon in Manitoba. *J. Mamm.* 45: 311-12.

Tevis, L., Jr. 1947. Summer activities of California raccoons. *J. Mamm.* 28: 323-32.

Whitney, L. F. 1933. The raccoon — some mental attributes. *J. Mamm.* 14: 108-14.

## Marten and fisher

Brander, R. B., and Books, D. J. 1973. Return of the fisher. *Nat. Hist.* 82(1): 52-57.

Brassard, J. A., and Bernard, R. 1939a. Observations on breeding and development of marten, *Martes americana* (Kerr). *Can. Field-Nat.* 53: 15-21.

Cowan, I. McT., and MacKay, R. H. 1950. Food habits of the marten *(Martes americana)* in the Rocky Mountain region of Canada. *Can. Field-Nat.* 64: 100-104.

Daniel, M. J. 1960. Porcupine quills in viscera of fisher. *J. Mamm.* 41: 133.

*de Vos, A. 1952. *The ecology and management of the fisher and marten in Ontario.* Ontario Dept. Lands and Forests Tech. Bull.

Eadie, W. R., and Hamilton, W. J., Jr. 1958. Reproduction in the fisher in New York. *N.Y. Fish Game J.* 5: 77-83.

Hamilton, W. J., Jr., and Cook. A. H. 1955. The biology and management of the fisher in New York. *N.Y. Fish Game J.* 2: 13-35.

Lensink, C. J.; Skoog, R. O.; and Buckley, J. L. 1955. Food habits of marten in interior Alaska and their significance. *J. Wildl. Mgmt.* 19: 364-68.

Markley, M. H. 1942. The breeding habits of marten. *Am. Fur Breeder* 14: 14-15.

Marshall, W. H. 1946. Winter food habits of the pine marten in Montana. *J. Mamm.* 27: 83-84.

———. 1951. Pine marten as a forest product. *J. For.* 49: 899-905.

Masters, R. D. 1980. Daytime resting sites of two Adirondack pine martens. *J. Mamm.* 61: 157.

Murie, A. 1961. Some food habits of the marten. *J. Mamm.* 42: 516-21.

Phillips, R. L., and Jonkel, C. eds. 1975. *Proceedings of the 1975 predator symposium.* Missoula, Montana: Montana Forest and Conservation Experiment Station.

Powell, R. A. 1978. A comparison of fisher and weasel hunting behavior. *Carnivore* 1: 28-34.

———. 1980. Fisher arboreal activity. *Can. Field-Nat.* 94: 90-91.

Quick, H. F. 1955. Food habits of marten *(Martes americana)* in northern British Columbia. *Can. Field.-Nat.* 69: 144-47.

———. 1956. Effects of exploitation on a marten population. *J. Wildl. Mgmt.* 20: 267-74.

Weckwerth, R. P., and Hawley, V. D. 1962. Marten food habits and population fluctuations in Montana. *J. Wildl. Mgmt.* 26: 55-74.

## Weasels

Boxall, P. C. 1979. Interaction between a long-tailed weasel and a snowy owl. *Can. Field-Nat.* 93: 67-68.

Brown, J. H., and Lasiewski, R. C. 1972. Metabolism of weasels: the cost of being long and thin. *Ecol.* 53: 939-43.

Burris, F. and Burris, D. 1974. Death of a cottontail. *Audubon.* 76: 36-37.

Byrne, A.; Stebbins, L. L.; and Delude, L. 1978. A new killing technique of the long-tailed weasel. *Acta Theriologica.* 23: 127-43.

*Criddle, N., and Criddle, S. 1925. The weasels of southern Manitoba. *Can. Field-Nat.* 29: 142-48.

Criddle, S. 1947. A nest of the least weasel. *Can. Field-Nat.* 61: 69.

Erlinge, S. 1975. Feeding habits of the weasel *Mustela nivalis* in relation to prey abundance. *Oikos* 26: 378-84.

Fitzgerald, B. M. 1977. Weasel predation on a cyclic population of the montane vole *(Microtus montanus)* in California. *J. Anim. Ecol.* 46: 367-97.

*Hall, E. R. 1951 *American weasels.* Kansas: University of Kansas Publs., Mus. Nat. Hist. pp. 87-96, 168-181, 193-222.

———. 1974. The graceful and rapacious weasel. *Nat. Hist.* 83(9): 44-50.

Hamilton, W. J., Jr. 1933. The weasels of New York. *Am. Midl. Nat.* 14: 289-344.

Heidt, G. A.; Peterson, M. K.; and Kirkland, G. L., Jr. 1968. Mating behavior and development of least weasels *(Mustela nivalis)* in captivity. *J. Mamm.* 49: 413-19.

Jeanne, R. L. 1965. A case of a weasel climbing trees. *J. Mamm.* 46: 344-45.

Lokemoen, J. T., and Higgins, K. F. 1972. Population irruption of the least weasel *(Mustela nivalis)* in east central North Dakota. *Prairie Nat.* 4: 96.

MacLean, S. F., Jr.; Fitzgerald, B. M.; and Pitelka, F. A. 1974. Population cycles in arctic lemmings: winter reproduction and predation by weasels. *Arctic Alp. Res.* 6: 1-12.

Northcott, T. H. 1971. Winter predation of *Mustela erminea* in Northern Canada. *Arctic* 24: 141-43.

Pearson, O. P. 1971. Additional measurements of the impact of carnivores on California voles *(Microtus californicus). J. Mamm.* 52: 41-49.

Polderboer, E. B. 1942. Habits of the least weasel *(Mustela rixosa)* in northeastern Iowa. *J. Mamm.* 23: 145-47.

Polderboer, E. B.; Kuhn, L. W.; and Hendrickson, G. O. 1941. Winter and spring habits of weasels in central Iowa. *J. Wildl. Mgmt.* 5: 115-19.

Powell, R. A. 1973. A model for raptor predation on weasels. *J. Mamm.* 54: 259-63.

Quick, H. F. 1944. Habits and economics of the New York weasel in Michigan. *J. Wildl. Mgmt.* 8: 71-78.

———. 1951. Notes on the ecology of weasels in Gunnison County, Colorado. *J. Mamm.* 32: 281-90.

Simms, D. A. 1979a. North American weasels: resource utilization and distribution. *Can. J. Zool.* 57: 504-20.

———. 1979b. Studies of an ermine population in southern Ontario. *Can. J. Zool.* 57: 824-32.

Teer, J. G. 1964. Predation by long-tailed weasels on eggs of blue-winged teal. *J. Wildl. Mgmt.* 28: 404-6.

Wright, P. L. 1947. The sexual cycle of the male long-tailed weasel *(Mustela frenata). J. Mamm.* 28: 343-52.

## Mink

Dunstone, N., and Sinclair, W. 1978. Comparative aerial and underwater visual acuity of the mink, *Mustela vison.* Schreber, as a function of discrimination distance and stimulus luminance. *Anim. Behav.* 26: 6-13.

Eberhardt, R. T. 1973. Some aspects of mink-waterfowl relationships on prairie wetlands. *Prairie Nat.* 5: 17-19.

Errington, P. L. 1961. *Muskrats and marsh management.* Harrisburg, Penn.: Stackpole.

———. 1963. *Muskrat populations.* Iowa: Iowa State University Press.

Hatler, D. F. 1976-77. The coastal mink on Vancouver Island, British Columbia. *Diss. Abs. Int.* B 37: 3296-97.

Marshall, W. H. 1935. Mink displays sliding habits. *J. Mamm.* 16: 228-29.

———. 1936. A study of the winter activities of the mink. *J. Mamm.* 17: 382-92.

Mitchell, J. L. 1961. Mink movements and populations on a Montana river. *J. Wildl. Mgmt.* 25: 48-54.

McCabe, R. A. 1949. Notes on live-trapping mink. *J. Mamm.* 30: 416-23.

Poole, T. B. 1978. An analysis of social play in polecats (Mustelidae) with comments on the form and evolutionary history of the open mouth play face. *Anim. Behav.* 26: 36-49.

Poole, T. B., and Dunstone, N. 1976. Underwater predatory behavior of the American mink *(Mustela vison). J. Zool.* 178: 395-412.

Schladweiler, J. L., and Storm, G. L. 1969. Den-use by mink. *J. Wildl. Mgmt.* 33: 1025-26.

Sealander, J. A. 1943. Winter food habits of mink in southern Michigan. *J. Wildl. Mgmt.* 7: 411-17.

Sinclair, W.; Dunstone, N.; and Poole, T. B. 1974. Aerial and underwater visual acuity in the mink *Mustela vison* Schreber. *Anim. Behav.* 22: 965-74.

Svihla, A. 1931. Habits of the Louisiana mink *(Mustela vison vulgivagus). J. Mamm.* 12: 366-68.

Yeager, L. E. 1943. Storing of muskrats and other foods by minks. *J. Mamm.* 24: 100-101.

## Wolverine

Beebe, W. 1940. Wolverines and men. *Bull. N.Y. Zool. Soc.* 43: 54-59.

Boles, B. K. 1977. Predation by wolves on wolverines. *Can. Field-Nat.* 91: 68-69.

Burkholder, B. L. 1962. Observations concerning wolverine. *J. Mamm.* 43: 263-64.

Grinnell, G. B. 1926. Some habits of the wolverine. *J. Mamm.* 7: 30-34.

Guiguet, C. J. 1951. An account of wolverine attacking mountain goat. *Can. Field-Nat.* 65: 187.

*Holbrow, W. C. 1976. *The biology, mythology, distribution, and management of the wolverine* (Gulo gulo) *in western Canada.* MNRM Practicum. Nat. Resour. Inst. Univ. Manitoba.

Koehler, G. M.; Hornocker, M. G.; and Hash, H. S. 1980. Wolverine marking behavior. *Can. Field-Nat.* 94: 339-41.

Krott, P. 1958. *Tupu-tupu-tupu.* London: Hutchinson.

———. 1960. Ways of the wolverine. *Nat. Hist.* 69(2): 16-29.

Quick, H. F. 1952. Some characteristics of wolverine fur. *J. Mamm.* 33: 492-93.

———. 1953. Wolverine, fisher, and marten studies in a wilderness region. *Trans. N.A. Wildl. Conf.* 18: 513-32.

Rausch, R. A., and Pearson, A. M. 1972. Notes on the wolverine in Alaska and the Yukon territory. *J. Wildl. Mgmt.* 36: 249-68.

van Zyll de Jong, C. G. 1975. The distribution and abundance of the wolverine (*Gulo gulo*) in Canada. *Can. Field-Nat.* 89: 431-37.

Wright, P. L., and Rausch, R. 1955. Reproduction in the wolverine, *Gulo gulo. J. Mamm.* 36: 346-55.

## Badger

Balph, D. F. 1961. Underground concealment as a method of predation. *J. Mamm.* 42: 423-24.

Cahalane, V. H. 1950. Badger coyote "partnerships." *J. Mamm.* 31: 354-55.

Curtis, W. 1978. Sure, it looks cute. . . . *Nat. Wildl.* 16(1): 37-39.

Dew, J. 1957. Badger's cold storage plant. *Blue Jay* 15: 177.

Drake, G. E., and Presnall, C. C. 1950. A badger preying upon carp. *J. Mamm.* 31: 355-56.

Drescher, H.-E. 1974. On the status of the badger, *Taxidea taxus,* in Manitoba (Canada). *Zool. Anz.* 192: 222-28.

Dobie, J. F. 1950. *The voice of the coyote.* London: Hammond, Hammond.

Errington, P. L. 1937. Summer food habits of the badger in northwestern Iowa. *J. Mamm.* 18: 213-16.

Knopf, F. L., and Balph, D. F. 1969. Badgers plug burrows to confine prey. *J. Mamm.* 50: 635-36.

Lindzey, F. G. 1976. Characteristics of the natal den of the badger. *Northwest Sci.* 50: 178-80.

———. 1978. Movement patterns of badgers in northwestern Utah. *J. Wildl. Mgmt.* 42: 418-22.

Long, C. A. 1973. *Taxidea taxus.* Mamm. Sp. No. 26.

Rathbun, A. P.; Wells, M. C.; and Bekoff, M. 1980. Cooperative predation by coyotes on badgers. *J. Mamm.* 61: 375-76.

Robinson, W. B., and Cummings, M. W. 1947. Notes on behavior of coyotes. *J. Mamm.* 28: 63-65.

Sargeant, A. B., and Warner, D. W. 1972. Movements and denning habits of a badger. *J. Mamm.* 53: 207-10.

Sawyer, E. J. 1925. Badger runs down ground squirrels. *J. Mamm.* 6: 125-26.

Snead, E., and Hendrickson, G. O. 1942. Food habits of the badger in Iowa. *J. Mamm.* 23: 380-91.

Stardom, R. P. 1978. Status report on American badger *Taxidea taxus* in Canada. Committee on the Status of Endangered Wildlife in Canada.

Wright, P. L. 1963. Variations in reproductive cycles in North American mustelids. In *Delayed implantation,* ed. A. C. Enders, pp. 77-98. Chicago: University of Chicago Press.

———. 1966. Observations on the reproductive cycles of the American badger (*Taxidea taxus*). *Symp. Zool. Soc. Lond.* 15:27-45.

## Skunks

Aleksiuk, M., and Stewart, A. P. 1977. Food intake, weight changes and activity of confined striped skunks *(Mephitis mephitis)* in winter. *Am. Midl. Nat.* 98: 331-42.

Allen, D. L. 1939. Winter habits of Michigan skunks. *J. Wildl. Mgmt.* 3: 212-28.

Bailey, T. N. 1971. Biology of striped skunks on a southwestern Lake Erie marsh. *Am. Midl. Nat.* 85: 196-207.

Crabb, W. G. 1948. The ecology and management of the prairie spotted skunk in Iowa. *Ecol. Monogr.* 18: 201-33.

Ewer, R. F. 1973. *The carnivores.* New York: Cornell University Press.

*Gunson, J. R., and Bjorge, R. R. 1979. Winter denning of the striped skunk in Alberta. *Can. Field-Nat.* 93: 252-58.

Houseknecht, C. R., and Tester, J. R. 1978. Denning habits of striped skunks *(Mephitis mephitis). Am. Midl. Nat.* 100: 424-30.

Mutch, G. R. P., and Aleksiuk, M. 1977. Ecological aspects of winter dormancy in the striped skunk *(Mephitis mephitis). Can. J. Zool.* 55: 607-15.

Schmidt, K. P. 1936. Dehairing of caterpillars by skunks. *J. Mamm.* 17: 287.

Stebler, A. M. 1938. Feeding behavior of a skunk. *J. Mamm.* 19: 374.

Storer, T. I., and Vansell, G. H. 1935. Bee-eating proclivities of the striped skunk. *J. Mamm.* 16: 118-21.

Storm, G. L. 1972. Daytime retreats and movements of skunks on farmlands in Illinois. *J. Wildl. Mgmt.* 36: 31-45.

Sunsquist, M. E. 1974. Winter activity of striped skunks *(Mephitis mephitis)* in east-central Minnesota. *Am. Midl. Nat.* 92: 434-46.

*Verts, B. J. 1967. *The biology of the striped skunk.* Urbana: University of Illinois Press.

Wade-Smith, J., and Richmond, M. E. 1978. Reproduction in captive striped skunks *(Mephitis mephitis). Am. Midl. Nat.* 100: 452-55.

## River otter

Field, R. J. 1970. Winter habits of the river otter *(Lutra canadensis)* in Michigan. *Papers Mich. Acad. Sci. Arts Lttrs.* 3: 49-58.

*Fur Production, Season 1977-78.* Ottawa: Statistics Canada. Cat. No. 23-207.

Greer, K. R. 1955. Yearly food habits of the river otter in the Thompson Lakes region, northwestern Montana, as indicated by scat analyses. *Am. Midl. Nat.* 54: 299-313.

Hamilton, W. J., Jr., and Eadie, W. R. 1964. Reproduction in the otter, *Lutra canadensis. J. Mamm.* 45: 242-52.

*Harris, C. J. 1968. *Otters, a study of the recent Lutrinae.* London: Weidenfeld and Nicolson.

Hooper, E. T., and Ostenson, B. T. 1949. Age groups in Michigan otter. *U. Mich. Occ. Papers. Mus. Zool.* 518: 1-22.

*Liers, E. E. 1951. Notes on the river otter *(Lutra canadensis). J. Mamm.* 32: 1-9.

Park, E. 1971. *The world of the otter.* New York: J. B. Lippincott.

Severinghaus, C. W., and Tanck, J. E. 1948. Speed and gait of an otter. *J. Mamm.* 29: 71.

Sheldon, W. G., and Toll, W. G. 1964. Feeding habits of the river otter in a reservoir in central Massachusetts. *J. Mamm.* 45: 449-54.

Toweill, D. E. 1974. Winter food habits of river otters in western Oregon. *J. Wild. Mgmt.* 38: 107-11.

## Sea otter

Bigg, M. A., and MacAskie, I. B. 1978. Sea otters reestablished in British Columbia. *J. Mamm.* 59: 874-76.

Costa, D. 1978. The sea otter: its interaction with man. *Oceanus* 21(2): 24-30.

Estes, J. A. 1980. *Enhydra lutris.* Mamm. Sp. 133.

Fisher, E. M. 1939. Habits of the southern sea otter. *J. Mamm.* 20: 21-36.

Hall, K. R. L., and Schaller, G. B. 1964. Tool-using behavior of the California sea otter. *J. Mamm.* 45: 287-98.

Hiscocks, B. 1977. Sea otters return to Canada's west coast. *Can. Geog. J.* 92(3): 20-27.

Houk, J. L., and Geibel, J. J. 1974. Observation of underwater tool use by the sea otter, *Enhydra lutris* Linnaeus. *Calif. Fish Game* 60: 207-8.

Kenyon, K. W. 1971. Return of the sea otter. *Nat. Geog.* 140: 520-39.

———. 1972. The sea otter. In *Mammals of the sea,* ed. S. H. Ridgway, pp. 205-14. Springfield, Ill.: Charles C. Thomas.

*———. 1975. *The sea otter in the eastern Pacific Ocean.* New York: Dover. Originally published in 1969 as *N. Am. Fauna* 68: 1-352.

Loughlin, T. R. 1980. Home range and territoriality of sea otters near Monterey, California. *J. Wildl. Mgmt.* 44: 576-82.

Morris, R.; Ellis, D. Y.; and Emerson, B. In press. The British Columbia transplant of sea otters *Enhydra lutris. Biological Conservation.*

Sandegren, F. E.; Chu, E. W.; and Vandevere, J. E. 1973. Maternal behavior in the California sea otter. *J. Mamm.* 54: 668-79.

Sherrod, S. K.; Estes, J. A.; and White, C. M. 1975. Depredation of sea otter pups by bald eagles at Amchitka Island, Alaska. *J. Mamm.* 56: 701-3.

Shimek, S. J. 1977. The underwater foraging habits of the sea otter, *Enhydra lutris. Calif. Fish Game* 63: 120-22.

Taylor, R. L., and Gough, B. 1977. New sighting of sea otter reported for Queen Charlotte Islands. *Syesis* 10: 177.

## Mountain lion

Dewar, P. 1976. British Columbia's shy cats. *Nature Can.* 5(4): 43-47.

Eaton, R. L., and Velander, K. A. 1977. Reproduction in the puma: biology, behavior and ontogeny. *The World's Cats* 3(3): 45-70.

Ewer, R. F. 1973. *The carnivores.* Ithaca: Cornell University Press.

Gashwiler, J. S., and Robinette, W. L. 1957. Accidental fatalities of the Utah cougar. *J. Mamm.* 38: 123-26.

Guggisberg, C. A. W. 1975. *Wild cats of the world.* New York: Taplinger Press.

Hornocker, M. G. 1969a. Stalking the mountain lion — to save him. *Nat. Geog.* 136: 638-55.

———. 1969b. Winter territoriality in mountain lions. *J. Wildl. Mgmt.* 33: 457-64.

*———. 1970. *An analysis of mountain lion predation upon mule deer and elk in the Idaho primitive area.* Wildl. Monogr. No. 21.

Nero, R. W., and Wrigley, R. E. 1977. Status and habits of the cougar in Manitoba. *Can. Field-Nat.* 91: 28-40.

Rabb, G. B. 1959. Reproductive and vocal behavior in captive pumas. *J. Mamm.* 40: 616-17.

Robinette, W. L.; Gashwiler, J. S.; and Morris, O. W. 1959. Food habits of the cougar in Utah and Nevada. *J. Wildl. Mgmt.* 23: 261-73.

———. 1961. Notes on cougar productivity and life history. *J. Mamm.* 42: 204-17.

*Seidensticker, J. C., IV; Hornocker, M. G.; Wiles, W. V.; and Messick, J. P. 1973. *Mountain lion social organization in the Idaho primitive area.* Wildl. Monogr. No. 35.

White, T. 1967. History of the cougar in Saskatchewan. *Blue Jay.* 25: 84-89.

Wright, B. S. 1973. *Cougar.* Ottawa: Canadian Wildlife Service Hinterlands Who's Who series.

## Bobcat and lynx

Bailey, T. N. 1972. The elusive bobcat. *Nat. Hist.* 81(8): 42-49.

*———. 1974. Social organization in a bobcat population. *J. Wildl. Mgmt.* 38: 435-46.

Barash, D. P. 1971. Cooperative hunting in the lynx. *J. Mamm.* 52: 180.

*Brand, C. J.; Keith, L. B.; and Fischer, C. A. 1976. Lynx responses to changing snowshoe hare densities in central Alberta. *J. Wildl. Mgmt.* 40: 416-28.

Brand, C. J., and Keith, L. B. 1979. Lynx demography during a snowshoe hare decline in Alberta. *J. Wildl. Mgmt.* 43: 827-49.

de Vos, A., and Matel, S. E. 1952. The status of the lynx in Canada, 1920-1952. *J. For.* 50: 742-45.

Elton, C., and Nicholson, M. 1942. The ten-year cycle in numbers of the lynx in Canada. *J. Anim. Ecol.* 11: 215-44.

Gashwiler, J. S.; Robinette, W. L.; and Morris, O. W. 1960. Foods of bobcats in Utah and eastern Nevada. *J. Wildl. Mgmt.* 24: 226-29.

Keith, L. B. 1977, *Lynx*. Ottawa: Canadian Wildlife Service Hinterland Who's Who series.

Nellis, C. H.; Wetmore, S. P.; and Keith, L. B. 1972. Lynx-prey interactions in central Alberta. *J. Wildl. Mgmt.* 36: 320-28.

Rollings, C. T. 1945. Habits, foods and parasites of the bobcat in Minnesota. *J. Wildl. Mgmt.* 9: 131-45.

Saunders, J. K., Jr. 1963a. Food habits of the lynx in Newfoundland. *J. Wildl. Mgmt.* 27: 384-90.

———. 1963b. Movements and activities of the lynx in Newfoundland. *J. Wildl. Mgmt.* 27: 390-400.

van Zyll de Jong, C. G. 1966. Food habits of the lynx in Alberta and the Mackenzie District, N.W.T. *Can. Field-Nat.* 80: 18-23.

Wolff, J. O. 1980. The role of habitat patchiness in the population dynamics of snowshoe hares. *Ecol. Monogr.* 50: 111-30.

## Introduction to sea lions and seals

Haley, D. 1978. Pinnipeds. In *Marine mammals of eastern North Pacific and Arctic waters,* ed. D. Haley, pp. 145-51. Seattle: Pacific Search Press.

Maxwell, G. 1967. *Seals of the world.* London: Constable.

Pike, G. C., and MacAskie, I. B. 1969. *Marine mammals of British Columbia.* Fish. Res. Bd. Can. Bull. 171.

Scheffer, V. B. 1958. *Seals, sea lions, and walruses: a review of the Pinnipedia.* Stanford, Calif.: Stanford University Press.

Stains, H. B. 1974. Carnivora. *Encyclopaedia Britannica Macropaedia* 3: 926-44.

## Sea lions

Farentinos, R. C. 1971. Some observations on the play behavior of the Steller sea lion *(Eumetopias jubata). Z. Tierpsych.* 28: 428-38.

Fiscus, C. H., and Baines, G. A. 1966. Food and feeding behavior of Steller and California sea lions. *J. Mamm.* 47: 195-200.

Gentry, R. L. 1974. The development of social behavior through play in the Steller sea lion. *Am. Zool.* 14: 391-403.

*Haley, D., ed. 1978. *Marine mammals of eastern North Pacific and Arctic waters.* Seattle: Pacific Search Press.

Harestad, A. 1977. Seasonal abundance of northern sea lions, *Eumetopias jubatus* (Schreber), at McInnes Island, British Columbia. *Syesis* 10: 173-74.

Harestad, A., and Fisher, H. D. 1975. Social behaviour in a non-pupping colony of Steller sea lions *(Eumetopias jubata) Can. J. Zool.* 53: 1596-1613.

Jameson, R. J., and Kenyon, K. W. 1977. Prey of sea lions in the Rogue River, Oregon. *J. Mamm.* 58: 672.

Kenyon, K. W., and Rice, D. W. 1961. Abundance and distribution of the Steller sea lion. *J. Mamm.* 42: 223-34.

Mathisen, O. A. 1959. Studies on Steller sea lion *(Eumetopias jubata)* in Alaska. *Trans. N. Am. Wildl. Conf.* 24: 346-56.

Mathisen, O. A.; Baade, R. T.; and Lopp, R. J. 1962. Breeding habits, growth and stomach contents of the Steller sea lion in Alaska. *J. Mamm.* 43: 469-77.

Orr, R. T., and Poulter, T. C. 1967. Some observations on reproduction, growth, and social behavior in the Steller sea lion. *Proc. Calif. Acad. Sci.* 35: 193-226.

Pike, G. C. 1958. Food of the northern sea lion. *Fish. Res. Bd. Can. Prog. Rep.* 112: 18-20.

*———. 1961. The northern sea lion in British Columbia. *Can. Aud.* 23(1): 1-5.

Pike, G. C., and MacAskie, I. B. 1969. *Marine mammals of British Columbia.* Fish. Res. Bd. Can. Bull. 171.

Pike, G. C., and Maxwell, B. E. 1958. The abundance and distribution of the northern sea lion *(Eumetopias jubata)* on the coast of British Columbia. *J. Fish. Res. Bd. Can.* 15: 5-17.

*Sandegren, F. 1975. Sexual-agonistic signalling and territoriality in the Steller sea lion *(Eumetopias jubatus). Rapp. P.-v. Réun. Cons. int. Explor. Mer* 169: 195-204.

Spalding, D. J. 1964. *Comparative feeding habits*

*of the fur seal, sea lion and harbour seal on the British Columbia coast.* Fish. Res. Bd. Can. Bull. 146.

## Harbor seal

Bigg, M. A. 1969. *The harbour seal in British Columbia.* Fish, Res. Bd. Can. Bull. 172.

Evans, W. E., and Bastian, J. 1969. Marine mammal communication: social and ecological factors. In *The biology of marine mammals,* ed. H. T. Andersen, pp. 425-75. New York: Academic Press.

Fisher, H. D. 1952. *The status of the harbor seal in British Columbia, with particular reference to the Skeena River.* Fish. Res. Bd. Can. Bull. 93.

Klinkhart, E. G. 1967. Birth of a harbor seal pup. *J. Mamm.* 48: 677.

*Knudtson, P. M. 1974. Birth of a harbor seal. *Nat. Hist.* 83(5): 30-37.

———. 1977. Observations of the breeding behavior of the harbor seal, in Humboldt Bay, California. *Calif. Fish Game* 63: 66-70.

Newby, T. C. 1973. Observations on the breeding behavior of the harbor seal in the state of Washington. *J. Mamm.* 54: 540-43.

*———. 1978. Pacific harbor seal. In *Marine mammals of eastern North Pacific and Arctic waters,* ed. D. Haley, pp. 184-91. Seattle: Pacific Search Press.

Pike, G. C., and MacAskie, I. B. 1969. *Marine mammals of British Columbia.* Fish. Res. Bd. Can. Bull. 171.

Scheffer, V. B., and Slipp, J. W. 1944. The harbor seal in Washington State. *Am. Midl. Nat.* 32: 373-416.

Spalding, D. J. 1964. *Comparative feeding habits of the fur seal, sea lion and harbour seal on the British Columbia coast.* Fish. Res. Bd. Can. Bull. 146.

Townsend, W. E., Jr. 1977. A harbor seal is born. *Pac. Disc.* 30(2): 24-27.

## Introduction to cloven-hoofed mammals

Carrington, R. 1963. *The mammals.* New York: Time Inc.

Church, D. C. 1969. *Digestive physiology and nutrition of ruminants,* vol. 1. Corvallis, Oregon: privately published.

Geist, V. 1971. *Mountain sheep — a study in behavior and evolution.* Chicago: University of Chicago Press.

Gentry, A. W. 1974. Artiodactyla. *Encyclopaedia Britannica Macropaedia* 2: 70-80.

Rue, L. L., III. 1978. *The deer of North America.* New York: Crown Publishers.

## Wapiti

Anderson, S., and Barlow, R. 1978. Taxonomic status of *Cervus elaphus merriami:* (Cervidae). *Southwest Nat.* 23: 63-70.

Altmann, M. 1952. Social behavior of elk, *Cervus canadensis nelsoni,* in the Jackson Hole area of Wyoming. *Behav.* 4: 116-43.

———. 1956. Patterns of herd behavior in free-ranging elk of Wyoming, *Cervus canadensis nelsoni. Zoologica* 41: 65-71.

———. 1960. The role of juvenile elk and moose in the social dynamics of their species. *Zoologica* 45: 35-39.

———. 1963. Naturalistic studies of maternal care in moose and elk. In *Maternal behavior in mammals,* ed. H. L. Rheingold, pp. 233-53. New York: John Wiley.

Boyce, M. S., and Hayden-Wing, L. D. 1979. *North American elk: ecology, behavior, and management.* University of Wyoming.

Craighead, J. J.; Craighead, F. C., Jr.; Ruff, R. L.; and O'Gara, B. W. 1973. *Home ranges and activity patterns of nonmigratory elk of the Madison drainage herd as determined by telemetry.* Wild. Monogr. No. 33.

de Vos, A.; Brokx, P.; and Geist, V. 1967. A review of social behavior of the North American cervids during the reproductive period. *Am. Midl. Nat.* 77: 390-417.

Flook, D. R. 1970. *Causes and implications of an observed sex differential in the survival of wapiti.* Canadian Wildlife Service Report Series No. 11.

Franklin, W. L.; Mossman, A. S.; and Dole, M. 1975. Social organization and home range of Roosevelt elk. *J. Mamm.* 56: 102-18.

Geist, V. 1966. The evolution of horn-like organs. *Behav.* 27: 175-213.

Knight, R. K. 1970. *The Sun River elk herd.* Wildl. Monogr. No. 23.

Kurtén, B. 1971. *The age of mammals.* London: Weidenfeld and Nicolson.

Murie, O. J. 1932. Elk calls. *J. Mamm.* 13: 331-36.

———. 1951. *The elk of North America.* Harrisburg, Penn.: Stackpole.

Shoesmith, M. W. 1978. Social organization of wapiti and woodland caribou. Ph.D. dissertation, University of Manitoba.

Struhsaker, T. T. 1967. Behavior of elk *(Cervus canadensis)* during the rut. *Zeit. Tierpsych.* 24: 80-114.

## Mule and white-tailed deer

de Vos. A.; Brokx, P.; and Geist, V. 1967. A review of social behavior of the North American cervids during the reproductive period. *Am. Midl. Nat.* 77: 390-417.

*Geist, V. 1966. The evolution of horn-like organs. *Behav.* 27: 175-213.

———. 1980. Downtown deer. *Nat. Hist.* 89(3): 56-65.

*Geist, V., and Walther, F., eds. 1974. *The social behavior of ungulates and its relation to management.* 2 vols. Morges, Switzerland: International Union for Conservation of Nature and Natural Resources.

Hirth, D. H. 1977. *Social behavior of white-tailed deer in relation to habitat.* Wildl. Monogr. No. 53.

Müller-Schwarze. D. 1971. Pheromones in the black-tailed deer, *(Odocoileus hemionus columbianus). Anim. Behav.* 19: 141-52.

Passmore, R. C. 1973. *White-tailed deer.* Ottawa: Canadian Wildlife Service Hinterland Who's Who series.

Prescott, W. H. 1974. Interrelationships of moose and deer of the genus *Odocoileus. Nat. Can.* 101: 493-504.

Rue, L. L. III. 1978. *The deer of North America.* New York: Crown Publishers.

Thomas, J. W.; Robinson, R. M.; and Marburger, R. G. 1965. Social behavior in a white-tailed deer herd containing hypogonadal males. *J. Mamm.* 46: 314-27.

## Moose

Altmann, M. 1956. Patterns of social behavior in big game. *Trans. N. Am. Wildl. Conf.* 21: 538-45.

———. 1958. Social integration of the moose calf. *Anim. Behav.* 6: 155-59.

*———. 1959. Group dynamics in Wyoming moose during the rutting season. *J. Mamm.* 40: 420-24.

———. 1963. Naturalistic studies of maternal care in moose and elk. In *Maternal behavior in mammals,* ed. H. L. Rheingold, pp. 233-53. New York: John Wiley.

Canadian Wildlife Service. 1973. *Moose.* Ottawa: Canadian Wildlife Service Hinterland Who's Who series.

Conley, J. D. 1956. Moose vs. bear. *Wyoming Wildl.* 20(9): 37.

de Vos, A.; Brokx, P.; and Geist, V. 1967. A review of social behavior of the North American cervids during the reproductive period. *Am. Midl. Nat.* 77: 390-417.

*Geist, V. 1963. On the behavior of the North American moose *(Alces alces andersoni* Patterson 1950) in British Columbia. *Behav.* 20: 377-416.

———. 1971. *Mountain sheep: a study in behavior and evolution.* Chicago: University of Chicago Press.

Geist, V., and Walther, F., eds. 1974. *The behavior of ungulates and its relation to management.* 2 vols. Morges, Switzerland: International Union for Conservation of Nature and Natural Resources.

Kelsall, J. P., and Prescott, W. 1971. *Moose and deer behavior in snow.* Canadian Wildlife Service Report Series No. 15.

Krefting, L. W. 1974. Moose distribution and habitat selection in north central North America. *Nat. Can.* 101: 81-100.

Lent, P. C. 1974. A review of rutting behavior in moose. *Nat. Can.* 101: 307-23.

Peterson, R. L. 1955. *North American moose.* Toronto: University of Toronto Press.

———. 1974. A review of the general life history of moose. *Nat. Can.* 101: 9-21.

Peterson, R. O., and Allen, D. L. 1974. Snow conditions as a parameter in moose-wolf relationships. *Nat. Can.* 101: 481-92.

*Stringham, S. F. 1974. Mother-infant relations in moose. *Nat. Can.* 101: 325-69.

## Caribou

Canadian Wildlife Service. 1973. *Caribou.* Canadian Wildlife Service Hinterland Who's Who series.

Dauphiné, T. C., Jr. 1976. *Biology of the Kaminuriak population of barren-ground caribou, part 4: growth, reproduction, and energy reserves.* Canadian Wildlife Service Report Series No. 38.

*Kelsall, J. P. 1968. *The migratory barren-ground caribou of Canada.* Ottawa: Department of Indian Affairs and Northern Development.

———. 1970. Migration of the barren-ground caribou. *Nat. Hist.* 49(7): 98-106.

Kelsall, J. P., and Klein, D. R. 1979. The state of knowledge of the Porcupine caribou herd. *Trans. N. Am. Wildl. Conf.* 44: 508-21.

Kelsall, J. P.; Telfer, E. S.; and Wright, T. D. 1977. *The effects of fire on the ecology of the boreal forest, with particular reference to the Canadian north: a review and selected bibliography.* Canadian Wildlife Service Occasional Paper No. 32.

Leopold, A. S., and Darling, F. F. 1953. *Wildlife in*

*Alaska: an ecological reconnaissance.* New York: Ronald Press.

Miller, D. R. 1976. *Biology of the Kaminuriak population of barren-ground caribou, part 3: taiga winter range relationships and diet.* Canadian Wildlife Service Report Series No. 36.

Miller, F. L. 1974. *Biology of the Kaminuriak population of barren-ground caribou, part 2: dentition as an indicator of age and sex; composition and socialization of the population.* Canadian Wildlife Service Report Series No. 31.

Miller, F. L., and Broughton, E. 1974. *Calf mortality on the calving ground of the Kaminuriak caribou, during 1970.* Canadian Wildlife Service Report Series No. 26.

Parker, G. R. 1972. *Biology of the Kaminuriak population of barren-ground caribou, part 1: total numbers, mortality, recruitment, and seasonal distribution.* Canadian Wildlife Service Report Series No. 21.

Symington, F. 1965. *Tuktu: a question of survival.* Ottawa: Canadian Wildlife Service.

Viereck, L. A. 1973. Wildfire in the taiga of Alaska. *Quat. Res.* 3: 465-95.

## Pronghorn

Autenrieth, R. E., and Fichter, E. 1975. *On the behavior and socialization of pronghorn fawns.* Wildl. Monogr. No. 42.

Bruns, E. H. 1977. Winter behavior of pronghorns in relation to habitat. *J. Wildl. Mgmt.* 41: 560-71.

Bullock, R. E. 1974. Functional analysis of locomotion in pronghorn antelope. In *The behavior of ungulates and its relation to management,* eds. V. Geist and F. Walther, vol. 1, pp. 274-305. Morges, Switzerland: International Union for Conservation of Nature and Natural Resources.

Einarson, A. S. 1948. *The pronghorn antelope.* Washington, D.C.: Wildlife Management Institute.

*Kitchen, D. W. 1974. *Social behavior and ecology of the pronghorn.* Wildl. Monogr. No. 38.

Mitchell, G. J., and Smoliak, S. 1971. Pronghorn antelope characteristics and food habits in Alberta. *J. Wildl. Mgmt.* 35: 238-50.

Webb, R.; Johnston, A.; and Soper, J. D. 1967. The prairie world. In *Alberta — a natural history,* ed. W. G. Hardy, pp. 93-116. Edmonton: Mismat Corp.

## Bison

Canadian Wildlife Service. 1974. *North American bison.* Ottawa: Canadian Wildlife Service Hinterland Who's Who series.

Fuller, W. A. 1960. Behaviour and social organization of the wild bison of Wood Buffalo National Park, Canada. *Arctic* 13: 3-19.

Haines, F. 1970. *The buffalo.* New York: Thomas Y. Crowell.

Kerr, D. G. G., and Davidson, R. I. K. 1966. *Canada — a visual history.* Toronto: Thomas Nelson and Sons.

Lott, D. F. 1972. Bison would rather breed than fight. *Nat. Hist.* 51(7): 40-45.

———. 1974. Sexual and aggressive behaviour of American bison *Bison bison.* In *The behaviour of ungulates and its relation to management,* eds. V. Geist and F. Walther, vol. 1, pp. 382-94. Morges, Switzerland: International Union for Conservation of Nature and Natural Resources.

McHugh, T. 1958. Social behavior of the American buffalo *(Bison bison bison). Zoologica* 43: 1-43.

———. 1972. *The time of the buffalo.* New York: Alfred A. Knopf.

Novakowski, N. S. 1978. Status report on wood bison *Bison bison athabascae* in Canada. Committee on the Status of Endangered Wildlife in Canada.

## Mountain goat

Chadwick, D. H. 1978. Daring guardians of the heights. *Nat. Geog.* 154: 284-96.

Geist, V. 1964. On the rutting behavior of the mountain goat. *J. Mamm.* 45: 551-68.

———. 1966. The evolution of horn-like organs. *Behav.* 27: 175-213.

———. 1967. On fighting injuries and dermal shields of mountain goats. *J. Wildl. Mgmt.* 31: 192-94.

———. 1971. *Mountain sheep — a study in behavior and evolution.* Chicago: University of Chicago Press.

———. 1971. Mountain goat behavior. *Wildl. Rev.* 5(10): 15-16.

Holroyd, J. C. 1967. Observations of rocky mountain goats on Mount Wardle, Kootenay National Park, B.C. *Can. Field-Nat.* 81: 1-22.

Samuel, W., and Macgregor, W. G., eds. 1977. *Proceedings of the first international mountain goat symposium.* British Columbia: Ministry of Recreation and Conservation, Fish and Wildlife Branch.

## Muskox

Gray, D. R. 1973. Social organization and behaviour of muskoxen *(Ovibos moschatus)* on Bathurst Island, N.W.T. Ph.D. dissertation, University of Alberta.

———. 1974. The defense formation of the musk-ox. *The Musk-ox* 14: 25-29.

———. 1975. *Muskox*. Ottawa: Canadian Wildlife Service Hinterland Who's Who series.

———. 1979. Movements and behavior of tagged muskoxen *(Ovibos moschatus)* on Bathurst Island, N.W.T. *Musk-ox* 25: 29-46.

Smith, P., and Jonkel, C. 1972. The return of the shaggy ox. *Nature Can.* 1(3): 20-21.

Tener, J. S. 1965. *Muskoxen in Canada*. Ottawa: Queen's Printer.

Wilkinson, P. F., and Shank, C. C. 1976. Rutting-fight mortality among musk oxen on Banks Island, Northwest Territories, Canada. *Anim. Behav.* 24: 756-58.

## Mountain sheep

Blood, D. 1973. *Mountain sheep*. Ottawa: Canadian Wildlife Service Hinderland Who's Who series.

*Geist, V. 1971. *Mountain Sheep — a study in behavior and evolution*. Chicago: University of Chicago Press.

## Picture Credits

Front Cover: Red fox by Arthur Savage
Back Cover: Mule deer by Wayne Lankinen
x Tom Brakefield
3 Craig Blacklock
7 Bristol Foster
9 Merlin D. Tuttle
10 Merlin D. Tuttle
11 Alvin E. Staffan
14 Beth Eldridge
15 Jeff Foott
17 Brian Milne
19 G. R. Parker
20 Stephen J. Krasemann/DRK Photo
21 Tom Brakefield
24 Vince Claerhout
27 Leonard Lee Rue III
29 Wayne Lankinen
30 Tom and Pat Leeson
32 Arthur Savage
33 Arthur Savage
34 Arthur Savage
35 Arthur Savage
36 Martin W. Grosnick
39 Wayne Lankinen
41 Jim Brandenburg
42 Martin W. Grosnick
44 Wolfgang Bayer
46 Tom W. Hall
47 Jeff Foott
50 Hälle Flygare
52 Alvin E. Staffan
55 Tim Fitzharris
57 Keith McDougall
58 Charles G. Summers, Jr.
61 Ken C. Balcomb
62 Ken C. Balcomb
65 Neil and Betty Johannsen
67 James Hudnall
68 John Ford
70 James Hudnall
71 Neil and Betty Johannsen
74 Wolfgang Bayer
76 Leonard Lee Rue III
77 Jim Brandenburg
79 Rollie Ostermick
81 Robert Garrott
82 Robert Garrott
85 Lynn Rogers
86 Arthur Savage
87 Brian Milne
89 Brian Milne
91 Charles G. Summers, Jr.
93 Stephen J. Krasemann/The Image Bank of Canada
95 Robert R. Taylor
97 Norman R. Lightfoot
98 Norman R. Lightfoot
101 Dennis Horwood
102 Tom W. Hall
104 Charles G. Summers, Jr.
105 Kent and Donna Dannen
106 Arthur Savage
108 Tom Brakefield
111 Stewart Cassidy
112 Tom Brakefield
113 Charles G. Summers, Jr.
115 Tom W. Hall
116 Stephen J. Krasemann/DRK Photo
119 Brian Milne
120 Tim Fitzharris
122 Jeff Foott
124 Jeff Foott
126 Tom W. Hall
127 Hälle Flygare
128 Martin W. Grosnick
129 Tom Brakefield
133 Neil and Betty Johannsen
135 Keith McDougall
137 Jeff Foott
138 Jeff Foott
141 Stephen J. Krasemann/DRK Photo
142 Jeff Foott
144 Norman R. Lightfoot
145 Charles G. Summers, Jr.
147 Jeff Foott
149 Doug Murphy
150 Stephen J. Krasemann/DRK Photo
151 Stephen J. Krasemann/DRK Photo
153 Brian Milne
154 Rollie Ostermick
156 Jeff Foott
157 Wilfried Schurig
158 Charles G. Summers, Jr.
160 Arthur Savage
161 Charles G. Summers, Jr.
163 Keith Gunnar
164 Craig Blacklock
167 Stephen J. Krasemann/DRK Photo
168 Philip S. Taylor
170 Leonard Lee Rue III
171 Stephen J. Krasemann/DRK Photo
172 Tom and Pat Leeson
173 Martin W. Grosnick

# Index


Age of Mammals, 2
Age of Reptiles, 2
Aggression. *See* Agonistic encounters
Agonistic encounters. *See also* Mating behavior; Ritualized aggression; Territories and territoriality
  in beaver, 4
  in bison, 161
  in black bears, 90
  in ground squirrels, 34
  in harbor seals, 138
  in hoary marmots, 28
  in meadow voles, 51
  in moose, 148, 150
  in mountain goats, 165
  in mountain sheep, 172, 174
  in muskoxen, 169
  in pronghorns, 156, 159
  in red foxes, 84
  in shrews, 6
  in woodchucks, 16
Alaska Coastal Indians, 165
*Alces alces.* *See* Moose
*Alopex lagopus.* *See* Arctic fox
American buffalo. *See* Bison
Amphibians, 1
*Antilocapra americana.* *See* Pronghorn
Antlers, 139, 141, 142, 143, 145-46, 147, 151, 152, 155, 156
*Aplodontia rufa.* *See* Mountain beaver
Aplodontidae. *See* Mountain beaver
Aquatic adaptations
  in beaver, 43
  in cetaceans, 59
  in muskrat, 53
  in pinnipeds, 131-32
Arctic fox, 80-83
Arctic ground squirrel, 3, 31, 34
Arctic hare, 19, 78, 83
Artiodactyla. *See* Cloven-hoofed mammals
Aztecs, 73

Badger, 33, 48, 73, 112-14, 117
Balaenidae. *See* Right whales
Balaenopteridae. *See* Rorquals
Baleen plates, 60, 66, 70
Baleen whales, 59, 60, 66, 69. *See also* Baleen plates; Blue whale; Bubble net; Gray whale; Humpback whale; Lunge feeding; Minke whale; Right whales; Rorquals
Barren-ground caribou, 152-55
Bats, 8-11. *See also* Flying fox; Smooth-faced bats; Vampire bats
  evolution of, 8
Beaked whales, 60
Bear trees, 88
Bearded seals, 83, 96, 131
Bears, 43, 73. *See also* Bear trees; Black bear; Grizzly bear; Polar bear
Bears, attacks on people, 90, 92-93
Beaver, 22, 23, 41-45, 53, 78, 110, 121
Beaver dams, 43
Beaver lodges, 42, 43, 109
Behavior, as naturally selected characteristic, 3, 4, 166
Beluga, 60, 96
"Bends," 59
Bering land bridge, 2, 169
Bighorn sheep, 139, 169-74
Birds, 2
Birth. *See* Parturition
Bison, 78, 92, 139, 155, 159, 159-62, 169, 174
*Bison bison.* *See* Bison
Black bear, 88-90, 151
"Blackfish," 60
Black-footed ferret, 73
Blacktail deer. *See* Mule deer
*Blarina brevicauda.* *See* Short-tailed shrew
Blowholes, 59
Blubber, 59, 123, 132
Blue whale, 59, 63, 69
Bobcat, 128-29, 130
Bovid family, 139, 155, 156. *See also* Bison; Mountain goat; Mountain sheep; Muskox
Breaching, 66, 69, 70
"Breathers," 54
Broadside display, 159, 165
Brown bear. *See* Grizzly bear
Bubble net, 70
Buffalo. *See* Bison
Buffalo birds, 161
Bugling, 143
Bushy-tailed wood rat, 49-50
Caecum, 22
California sea lion, 131, 132
Calls. *See* Vocalization
Camel, 3, 139
Camouflage, 19, 80, 103, 138
Cannibalism, 34-35, 83, 88
Canine teeth. *See* Mammals, dentition of
Canidae. *See* Dog family
Canids. *See* Dog family
*Canis latrans.* *See* Coyote
*Canis lupus.* *See* Wolf
"Caravaning," 6
Caribou, 78, 83, 92, 110, 139, 143, 152-55
Carnassial teeth, 72, 73, 78, 110
Carnivora. *See* Carnivores
Carnivorean lethargy, 96, 99, 117
Carnivores, 19, 50, 72-130, 131, 136. *See also* Bears; Cat family; Dog family; Raccoon; Weasel family
  dentition of, 72, 73, 78, 94, 110
  evolution of, 72, 73
  taxonomy of, 72-73
Cascade golden-mantled ground squirrel, 31, 35
*Castor canadensis.* *See* Beaver
Castoreum, 43, 45
Castoridae. *See* Beaver
*Castoroides ohioensis,* 45
Cat family, 72, 73. *See also* Bobcat; Cats, domesticated; Lynx; Mountain lion; Saber-toothed cats
Cats, domesticated, 73
Cattle, 139
Cave bats, 10
Cellulose, digestion of, 13, 22, 139-40
Cervids. *See* Deer family
*Cervus elaphus.* *See* Wapiti
Cetaceans, 59-71. *See also* Dolphins; Porpoises; Whales
  taxonomy of, 59-60
Chipmunks, 23-25, 27, 35, 106. *See also* Least chipmunk; Yellow-pine chipmunk
Chiroptera. *See* Bats
Class, 3
Classification. *See* Taxonomy

*Clethrionomys gapperi. See* Gapper's red-backed vole
Cloven-hoofed mammals, 13, 75, 110, 139-74. *See also* Bovid family; Deer family; Pronghorn
evolution of, 140, 169
taxonomy of, 139
Coati, 97
Cold-blooded animals, 1. *See also* Amphibians; Reptiles
Collared pika, 15
Columbian ground squirrel, 16, 31, 33, 34, 114
Common seal. *See* Harbor seal
Communal denning, 86
Communication. *See* Vocalization
Competitive exclusion, 25
*Condylura cristata. See* Star-nosed mole
Conservation and endangered species, 4, 28, 43, 68, 69, 73, 94, 96, 100, 111, 114, 124, 125, 138, 155, 162, 169
Conservation of body heat. *See* Thermoregulation
Convergent evolution, 13, 132
Copulation, 58, 66
Cottontails, 6, 13, 16-19, 27, 35, 75, 88, 104, 107, 114, 130. *See also* Eastern cottontail; Nuttall's cottontail
Cougar. *See* Mountain lion
Courtship. *See* Mating behavior
Cows. *See* Cattle
Coyolxauhqui, 73
Coyote, 16, 21, 33, 38, 43, 51, 73-76, 88, 100, 114, 155, 159, 165, 172
Creodonts, 72
Cricetidae. *See* Mice; Muskrat; Rats; Voles
Cross fox, 84
Cud-chewing. *See* Ruminants and rumination

Dall's porpoise, 64, 65
Dall's sheep, 169, 172
Deer, 56, 75, 78, 103, 110, 125. *See also* Mule deer; White-tailed deer
Deer family, 139, 155, 159, 165. *See also* Caribou; Moose; Mule deer; Wapiti; White-tailed deer
Deer mouse, 45-48, 84
Delayed implantation, 100, 114
Delphinidae. *See* Dolphins; Killer whale; Porpoises
Dentition. *See* Mammals, dentition of
Dewclaws, 139, 154
Diaphragm, 2
Diastema, 22
Digestion in herbivores. *See* Cellulose, digestion of
Dinosaurs, 2
Dog family, 72-73. *See also* Coyote; Dogs, domesticated; Foxes; Wolf
Dogs, domesticated, 19, 33, 77
Dolphins, 8, 59, 60, 64. *See also* Pacific white-sided dolphin
Dominance order
defined, 4
in bison, 161
in cottontails, 18-19
in coyotes, 75-76
in deer, 146
in deer mice, 46, 48
in mountain goats, 162, 165
in mountain sheep, 172, 174
in pronghorns, 156, 159
in red foxes, 84
in wapiti, 140
in wolves, 77, 78
Douglas's squirrel, 37
Dual-action jaw, 22

Eared seals, 131. *See also* Northern fur seal; Sea lions
Earless seals, 131. *See also* Bearded seal; Harbor seal; Harp seal; Northern elephant seal; Ringed seal
Eastern cottontail, 16, 18-19
Echolocation
in bats, 8-9
in killer whales, 61
in porpoises, 65
in shrews, 7
in toothed whales, 60
Elk. *See* Wapiti
Endangered species. *See* Conservation and endangered species
Endothermy, 1-2. *See also* Thermoregulation
*Enhydra lutris. See* Sea otter
*Erethizon dorsatum. See* Porcupine
Erethizontidae. *See* Porcupine
Ermine, 16, 103-107
Eschrichtiidae. *See* Gray whale
*Eschrichtius robustus. See* Gray whale
*Eumetopias jubatus. See* Steller sea lion
*Eutamias amoenus. See* Yellow-pine chipmunk
*Eutamias minimus. See* Least chipmunk
Evolution. *See* Convergent evolution; Mammals, evolution of; Natural selection
Extinctions, Pleistocene, 2-3
Extinctions, recent, 73, 94

Family, 3
"Feeders," 53-54
Felidae. *See* Cat family
Felids. *See* Cat family
*Felis concolor. See* Mountain lion
*Felis lynx. See* Lynx
*Felis rufus. See* Bobcat
Field mouse. *See* Meadow vole
Fighting. *See* Agonistic encounters
Fin whale, 69
"Finners." *See* Rorquals
Fire, 151, 152, 161
Fisher, 58, 73, 100, 102-103
Fluke standing, 66
Fluking up, 67
Flying fox, 8
Flying squirrels, 23, 100. *See also* Northern flying squirrel
Forest fires. *See* Fire
"Forms," 21
Foxes, 19, 33, 48, 51, 78, 99. *See also* Arctic fox; Red fox; Swift fox
Franklin's ground squirrel, 31, 33, 35, 114

Gapper's red-backed vole, 51, 100
Genus, 3-4
Geomyidae. *See* Pocket gophers
Giant armadillo, 3
Giant beaver, 45
Giraffe, 139
*Glaucomys sabrinus. See* Northern flying squirrel
Gliding membranes, 38, 39

"Glutton, " 110
Goats, 139, 169. *See also* Mountain goat
Golden-mantled ground squirrel, 31, 35
Gophers. *See* Ground squirrels; Pocket gophers; Striped gopher
Gopher cores, 23
Grant's caribou, 152, 154
Gray matter, 2
Gray whale, 60, 63, 65, 66-68
Gray wolf. *See* Wolf
Great horned owl, 21, 117
Grizzly bear, 28, 83, 88, 89, 90-94, 95, 117, 165
Ground sloth, 3
Ground squirrels, 23, 30, 31-35, 92, 100, 112-14. *See also* Arctic ground squirrel; Cascade golden-mantled ground squirrel; Columbian ground squirrel; Franklin's ground squirrel; Golden-mantled ground squirrel; Richardson's ground squirrel; Thirteen-lined ground squirrel
Groundhog, 26
*Gulo gulo*. *See* Wolverine

Habitat. *See* Life zones; Microhabitat
Hair, 2, 59
Harbor porpoise, 64, 65
Harbor seal, 63, 131, 136-38
Harems, 134, 142, 169
Hares, 13, 16, 19-21, 56, 75, 104. *See also* Arctic hare; Snowshoe hare; White-tailed jack rabbit
Harp seal, 96, 131
Hawks. *See* Raptors
Heart, four-chambered, 2
Hedgehogs, 5
Herbivores. *See* Cellulose, digestion of
Heteromyidae. *See* Pocket mice; Kangaroo rats
Hibernation
  in bats, 10, 12
  in black bears, 90
  in chipmunks, 25
  in ground squirrels, 90
  in Richardson's ground squirrel, 31, 33
  in woodchucks, 26
Hippopotamus, 139
Hoary bat, 8
Hoary marmot, 16, 28, 30
Home range, defined, 4
*Homo sapiens*. *See* Human beings
Hooves, 139, 145, 148, 154, 159, 162, 166
Horns, 139, 155, 156, 158, 159, 161, 165, 169, 170, 172, 174
Horse, 3, 139
Howling, 76, 80
Human beings, 1, 2, 3 4, 12, 19, 22, 77
Humpback whale, 4, 60, 63, 69-71
Hyena, 72

Ice Ages, 2
Incisors. *See* Dentition
"Indian devil," 110
Indians and Inuit, 2-3, 6, 73, 76, 161, 165, 166
Insectivora. *See* Insectivores
Insectivores, 5-7, 8. *See also* Hedgehogs; Moles; Shrews
  evolution of, 5
Intelligence. *See* Mammals, intelligence of
Inter-digital glands, 146
Inuit, 6, 166

Jack rabbits, 130. *See also* White-tailed jack rabbit
Jumping deer, 145
Jumping mice, 23

Kangaroo rats, 23
Keratin, 139
Killer whale, 59-60, 60-63, 67, 68, 124, 134, 138
Kinkajou, 97

*Lagenorhnychus obliquidens*. *See* Pacific white-sided dolphin
Lagomorpha. *See* Lagomorphs
Lagomorphs, 13-21, 75, 129. *See also* Cottontails; Hares; Pikas
  dentition of, 13
  evolution of, 13
  taxonomy of, 13
*Lasiurus cinereus*. *See* Hoary bat
Learning, 2, 121
Least chipmunk, 24, 25
Least weasel, 104, 106
Leg-hold trap, 111
Lemmings, 83, 88, 92, 106
Leporids. *See* Cottontails; Hares
*Lepus americanus*. *See* Snowshoe hare
*Lepus arcticus*. *See* Arctic hare
*Lepus townsendii*. *See* White-tailed jack rabbit
Life zones, 1
Linnaeus. *See* Linné, Karl von
Linné, Karl von, 97
Little brown bat, 8, 10, 12
Lobtailing, 69
"Lone wolf," 78, 80
Long-tailed weasel, 16, 103-7
"Lumpers," 90
Lunge feeding, 70
*Lutra canadensis*. *See* River otter
Lynx, 20, 38, 128-30, 165

Mammalia, 3. *See also* Mammals, characteristics of
Mammals, characteristics of, 1-2
Mammals, dentition of, 2. *See also* Carnivores, dentition of; Lagomorphs, dentition of; Porpoises, dentition of; Rodents, dentition of; Sea lions, dentition of; Walrus, dentition of
Mammals, evolution of, 2, 56. *See also* Bats, evolution of; Carnivores, evolution of; Cloven-hoofed mammals, evolution of; Insectivores, evolution of; Lagomorphs, evolution of; Pinnipeds, evolution of
Mammals, intelligence of, 2, 84
Mammals, taxonomy of, 3. *See also* Carnivores, taxonomy of; Cetaceans, taxonomy of; Cloven-hoofed mammals, taxonomy of; Lagomorphs, taxonomy of; Pinnipeds, taxonomy of; Rodents, taxonomy of
Mammary glands, 2
Man. *See* Human beings
Marine adaptations. *See* Aquatic adaptations

*Marmota caligata. See* Hoary marmot
*Marmota flaviventris. See* Yellow-bellied marmot
*Marmota monax. See* Woodchuck
*Marmota vancouverensis. See* Vancouver Island marmot
Marmots, 23 28-30, 92, 110. *See also* Hoary marmot; Vancouver Island marmot; Yellow-bellied marmot
Marten, 16, 38, 73, 100-103, 110, 114
*Martes americana. See* Marten
*Martes pennanti. See* Fisher
Masked shrew, 5, 6, 7
Maternal behavior. *See* Parental care
Mating behavior
  in bison, 161
  in cottontails, 18
  in coyotes, 75-76
  in deer, 146
  in deer mice, 46
  in gray whales, 66
  in humpback whales, 69
  in moose, 148, 150
  in mountain goats, 165
  in mountain lions, 125
  in mountain sheep, 172
  in muskoxen, 169
  in porcupines, 56, 58
  in pronghorns, 156
  in snowshoe hares, 21
  in Steller sea lions, 134
  in wapiti, 142
Meadow vole, 5, 6, 51-53, 106, 107
*Megaptera novaenangliae. See* Humpback whale
*Mephitis mephitis. See* Striped skunk
Metabolic rate. *See also* Carnivorean lethargy; Hibernation; Thermoregulation
  in porpoises, 64-65
  in sea otters, 123
  in shrews, 5-6
Miacids, 72, 73
Mice, 22, 23, 27, 51, 75, 78, 83, 88, 92, 100, 103, 106, 110, 114. *See also* Deer mouse; Jumping mice; Pocket mice; Voles
Microhabitat, 1
*Microsorex hoyi. See* Pygmy shrew
*Microtus pennsylvanicus. See* Meadow vole
Migration
  in arctic foxes, 83
  in bats, 10, 12
  in caribou, 152, 153
  in gray whales, 66, 68
  in humpback whales, 69
  in mule deer, 146
  in polar bears, 96
  in wapiti, 140
Mink, 54, 73, 107-9
Minke whale, 63, 69
Molars. *See* Dentition
Moles, 5. *See also* Star-nosed mole
Mongoose, 72
Monodontidae. *See* Beluga; Narwhal
Moose, 78, 92, 103, 110, 139, 143, 148-51, 172
Mountain beaver, 23
Mountain cottontail. *See* Nuttall's cottontail
Mountain devil, 127
Mountain goat, 110, 139, 162-66, 169, 174
Mountain lion, 1, 58, 111, 125-27, 165, 170
Mountain sheep, 92, 165, 169-74. *See also* Bighorn sheep; Thinhorn sheep
Mule deer, 78, 130, 139, 143-47
Muskox, 78, 110, 139, 166-69
Muskrat, 23, 53-55, 99, 103, 107, 109, 121
*Mustela erminea. See* Ermine
*Mustela frenata. See* Long-tailed weasel
*Mustela nivalis. See* Least weasel
*Mustela vison. See* Mink
Mustelidae. *See* Weasel family
Mustelids. *See* Weasel family
Myoglobin, 59
*Myotis lucifugus. See* Little brown bat
Mysticeti. *See* Baleen whales

Narwhal, 60, 96
Natural selection, 4
*Neotoma cinerea. See* Bushy-tailed wood rat
Niche, defined, 4
Northern elephant seal, 131
Northern flying squirrel, 38-40, 103
Northern fur seal, 131
Northern sea lion. *See* Steller sea lion
Norway rat, 50
Nursing. *See also* Mammary glands
  in arctic foxes, 80
  in bats, 12
  in beaver, 43, 44
  in bushy-tailed wood rats, 49
  in cottontails, 17
  in coyotes, 76
  in harbor seals, 136
  in moose, 148
  in muskoxen, 166, 169
Nuttall's cottontail, 16

*Ochotona collaris. See* Collared pika
*Ochotona princeps. See* Rocky Mountain pika
Ochotonidae. *See* Pikas
Odd-toed ungulates, 139
Odobenidae. *See* Walrus
*Odobenus rosmarus. See* Walrus
*Odocoileus hemionus. See* Mule deer
*Odocoileus virginianus. See* White-tailed deer
Odontoceti. *See* Toothed whales
*Ondatra zibethicus. See* Muskrat
*Omingmak,* 166
Opossum, 56
*Orcinus orca. See* Killer whale
Order, 3
*Oreamnos americanus. See* Mountain goat
Otariidae. *See* Eared seals
Otters, 73. *See also* River otter; Sea otter
Overhunting, 4, 93-94, 125, 134, 138, 140, 152, 155, 162, 166, 169. *See also* Conservation and endangered species; Trappers and trapping
*Ovibos moschatus. See* Muskox
*Ovis canadensis. See* Bighorn sheep
*Ovis dalli. See* Thinhorn sheep
Owls. *See* Raptors

Pacific white-sided dolphin, 64
Pack rat, 49
Paleo-Indians, 2-3

Panda, 97
Parental care, 2. *See also* Nursing; Parturition; Paternal care
in arctic foxes, 80
in bats, 12
in black bears, 88-89
in bushy-tailed wood rats, 49
in coyotes, 76
in deer, 146
in deer mice, 48
in grizzly bears, 92
in harbor seals, 136
in humpback whales, 69, 70
in moose, 148, 149, 150-51
in mountain goats, 163, 165
in mountain lions, 126
in mountain sheep, 170
in muskrats, 54
in northern flying squirrels, 40
in pikas, 16
in polar bears, 96
in pronghorns, 159
in red foxes, 86
in river otters, 121
in sea otters, 124
in shrews, 6
in Steller sea lions, 134
in striped skunks, 117
in wolverines, 110
Parturition
in cottontails, 18
in harbor seals, 136
in killer whales, 63
Paternal care
in arctic foxes, 80
in coyotes, 76
in deer mice, 48
in red foxes, 86
Peary caribou, 154
Peck orders. *See* Dominance hierarchies
People. *See* Human beings
Perissodactyls. *See* Odd-toed ungulates
*Peromyscus maniculatus. See* Deer mouse
*Phoca vitulina, See* Harbor seal
Phocidae. *See* Earless seals
*Phocoena phocoena. See* Harbor porpoise
*Phocoenoides dalli. See* Dall's porpoise
Physeteridae. *See* Sperm whales
Pigs, 139
Pikas, 13-16, 30, 100. *See also* Collared pika; Rocky mountain pika
Pinnipedia. *See* Pinnipeds
Pinnipeds. 131-38. *See also* Eared seals; Earless seals; Walrus
evolution of, 131, 132, 136
taxonomy of, 131
Pipeline construction 94
Plains bison, 159-62
Plains grizzly, 94
Plains Indians, 161
Play
in black bears, 90
in hoary marmots, 28
in mink, 109
in mountain goats, 165
in pronghorns, 159
in red foxes, 84
in red squirrels, 38
in river otters, 118
in wapiti, 140
in wolves, 78
Pleistocene extinctions, 2-3
Pocket gophers, 23, 106, 114
Pocket mice, 23
Poisonous saliva, 6
Polar bear, 73, 83, 94-96
Pollution, 96, 138
Polychlorinated biphenyls (PCBs), 96, 138
Population cycles
in arctic foxes, 83
in coyotes, 20
in foxes, 20, 88
in lemmings, 83, 106
in lynx, 20, 128, 130
in meadow voles, 51, 53, 106
in raptors, 20
in snowshoe hares, 20-21, 128, 130
in weasels, 106
Porcupine, 23, 56-58, 103, 111, 125
Porpoises, 8, 59, 60, 64. *See also* Dall's porpoise; Harbor porpoise
dentition of, 65
Predator-prey relations, 19, 78, 106. *See also* Population cycles
Premolars. *See* Dentition
Primates, 22
*Procyon lotor. See* Raccoon
Procyonidae. *See* Coati; Kinkajou; Panda; Raccoon
Pronghorn, 139, 155-59, 165, 172
Puffing pig, 64
"Push-ups," 54
Pygmy shrew, 5

Quill pig, 56
Quills, 57, 58

Rabbits. *See* Cottontails; White-tailed jack rabbit
Rabies, 12, 83, 117
Raccoon, 27, 48, 73, 97-99
"Radar." *See* Echolocation
Rallying stations, 86
*Rangifer tarandus. See* Caribou
*Rangifer tarandus groenlandicus. See* Barren ground caribou
Raptors, 16, 19, 21, 28, 33, 38, 40, 48, 50, 51, 99, 107, 117, 170. *See also* Great horned owl
Rats, 23, 53, 106. *See also* Bushy-tailed wood rat; Kangaroo rats; Norway rat
Red blood cells, 59
Red fox, 21, 27, 84-88, 107
Red squirrel, 36-38, 40, 100, 103
Reindeer, 152
Reingestion, 13, 22
Rendezvous sites, 80
Reptiles, 1, 2
Richardson's ground squirrel, 31, 33-34, 35
Right whales, 60
Ringed seals, 83, 94, 96, 131
Ritualized aggression, 4
in bison, 161
in black bears, 88
in moose, 148
in mountain goats, 165
in red foxes, 84
in Steller sea lions, 134
in wapiti, 142-43
in white-tailed deer, 146
River otter, 27, 107, 109, 118-21, 121
Rockchuck, 28
"Rock rabbit," 15
Rocky Mountain pika, 13, 15
Rocky Mountain wapiti, 140
Rodentia. *See* Rodents
Rodents, 13, 22-58, 75, 92, 106, 112, 130. *See also* Beaver; Mice; Mountain beaver; Muskrat; Pocket gophers; Porcupine; Rats; Squirrels; Voles

dentition of, 22, 41
taxonomy of, 23
Roosevelt elk, 140
Rorquals, 60, 69, 70. *See also* Humpback whale
Ruminants and rumination, 13, 139-40
Rutting pits, 165

Saber-toothed cats, 3
Scent-bomb defense, 73, 110, 115
Scent-marking, 4
in beaver, 45
in bobcats and lynx, 130
in deer, 146
in mountain lions, 125
in pikas, 16
in pronghorns, 159
in red foxes, 84, 88
in shrews, 6
in wapiti, 143
in wolves, 80
Sciuridae. *See* Squirrels
Scrapes, 125, 143, 146
Sea lions, 60, 63, 131, 132-35, 136. *See also* California sea lion; Steller sea lion
dentition of, 134
Sea mammals, 83. *See also* Cetaceans; Pinnipeds
Sea otter, 4, 73, 136
Sea serpent, 118
Seals, 131. *See also* Eared seals; Earless seals
Sei whale, 69
Sheep, 75, 139, 169. *See also* Mountain sheep
Short-tailed shrew, 5, 6
Short-tailed weasel. *See* Ermine
Shrews, 5, 8, 103, 106. *See also* Masked shrew; Pygmy shrew; Short-tailed shrew
Silver fox, 84
"Skunk-bear," 110
Skunks, 27, 73, 99, 115-17. *See also* Striped skunk; Western spotted skunk
Smoky the Bear, 151
Smooth-faced bats, 8. *See also* Cave bats; Hoary bat; Little brown bat; Tree bats
Sneak cat, 127
Snowshoe hare, 19-21, 88, 100, 103, 104, 110, 128, 130
Snowshoe hare cycle. *See* Population cycles
Sociability, adaptive value of
in caribou, 155
in hoary marmots, 28, 30
in muskoxen, 169
in pronghorns, 155-56
Social behavior, defined, 4
Sonar. *See* Echolocation
Songs. *See* Vocalization
*Sorex cinereus. See* Masked shrew
Species, defined, 3
Sperm whales, 59, 60
*Spermophilus columbianus. See* Columbian ground squirrel
*Spermophilus franklinii. See* Franklin's ground squirrel
*Spermophilus lateralis. See* Golden-mantled ground squirrel
*Spermophilus parryii. See* Arctic ground squirrel
*Spermophilus richardsonii. See* Richardson's ground squirrel
*Spermophilus saturatus. See* Cascade golden-mantled ground squirrel
*Spermophilus tridecemlineatus. See* Thirteen-lined ground squirrel
*Spilogale gracilis. See* Western spotted skunk
"Splitters," 90
Spotted skunk. *See* Western spotted skunk
Spyhopping, 62, 63, 66
Squirrels and squirrel family, 23, 27, 40, 110. *See also* Chipmunks; Ground squirrels; Marmots; Northern flying squirrel; Red squirrel
Standing-over, 75
Star-nosed mole, 5
Steller sea lion, 131, 132-35
Stoat. *See* Ermine
Stomach, four-chambered, 139
Stomach, three-chambered, 65
Stone's sheep, 169-70
Striped gopher, 33
Striped skunk, 115-17
Subspecies, defined, 4
Surface-to-volume ratio, 5, 59, 166
Sweat glands, 2
Swan Hills grizzlies, 94
Swift fox, 73
*Sylvilagus floridanus. See* Eastern cottontail
*Sylvilagus nuttallii. See* Nuttall's cottontail

*Tamiasciurus douglasii. See* Douglas's squirrel
*Tamiasciurus hudsonicus. See* Red squirrel
Tapir, 3
Tarsal glands, 146
*Taxidea taxus. See* Badger
Taxonomy, 3. *See also* Linné, Karl von; "Lumpers"; Mammals, taxonomy of; "Splitters"
Teeth. *See* Dentition
Territories and territoriality
in beaver, 45
in bushy-tailed wood rats, 50
in coyotes, 76
in deer mice, 48
in ground squirrels, 37, 38
in humpback whales, 69
in mountain lions, 125, 126
in pikas, 16
in pronghorns, 159
in red foxes, 84
in red squirrels, 37, 38
in sea otters, 124
in shrews, 6
in Steller sea lions, 134
in wolves, 78, 80
Territory, defined, 4
Therapsids, 2
Thermoregulation, 1, 2
in cetaceans, 59
in muskoxen, 166
in pinnipeds, 132
in porpoises, 64
in sea otters, 123
in shrews, 5-6
in weasels, 107
Thinhorn sheep, 139, 169-74. *See also* Dall's sheep; Stone's sheep
Thirteen-lined ground squirrel, 31, 34, 35, 114
Throat grooves, 60, 70
Tiger moths, 10
Timber wolf. *See* Wolf
Tool use, 123
Toothed whales, 59-60. *See also* Beaked whales; Beluga; Dall's porpoise; Dolphins; Echolocation; Harbor porpoise; Killer whale; Narwhal; Porpoises; Sperm whales

Trade rat, 49
Trappers and trapping, 4, 40, 43, 54, 73, 83, 103, 110, 110-11, 121, 123-24, 130
Tree bats, 10
Tree squirrels, 37. *See also* Douglas's squirrel; Northern flying squirrel; Red squirrel
Tusks, 131

Ungulates. *See* Cloven-hoofed mammals; Odd-toed ungulates
*Ursus americanus. See* Black bear
*Urus arctos. See* Grizzly bear
*Ursus lotor. See* Raccoon
*Ursus maritimus. See* Polar bear
Ursidae. *See* Bears

Vampire bat, 8
Vancouver Island marmot, 4, 28
Varying hare. *See* Snowshoe hare
"Velvet," 141, 143, 145, 151, 152
Vespertilionidae. *See* Smooth-faced bats
Vocalization
  in chipmunks, 23
  in coyotes, 76
  in Douglas's and red squirrels, 37
  in hoary marmots, 28
  in humpback whales, 69
  in killer whales, 61
  in moose, 148
  in pikas, 13, 16
  in porcupines, 58
  in porpoises, 65
  in red foxes, 84
  in sea lions, 132
  in shrews, 6
  in wapiti, 143
  in wolves, 80
  in woodchucks, 27
Voles, 23, 27, 51, 83, 88, 100, 106, 110, 114, 130. *See also* Gapper's red-backed vole; Meadow vole
Von Linné, Karl. *See* Linné, Karl von
*Vulpes vulpes. See* Red fox

Wallows, 150, 161, 162
Walrus, 63, 131
  dentition of, 131
Wapiti, 75, 78, 89, 92, 110, 125, 139, 140-43
Warmbloodedness. *See* Thermoregulation
Weaning. *See* Nursing
Weasel family, 73, 111, 114. *See also* Badger; Black-footed ferret; Fisher; Marten; Mink; Otters; Skunks; Weasels; Wolverine
Weasels, 19, 33, 48, 73, 103-7, 121. *See also* Ermine; Least weasel; Long-tailed weasel
Western spotted skunk, 115-16
Whales, 8, 59, 110. *See also* Aquatic adaptations; Baleen whales; Blowholes; Breaching; Fluke standing; Lobtailing; Spyhopping; Toothed whales
Whalers and whaling, 4, 63, 68
"Whirling," 165
Whistle-pig, 27
"Whistler," 30
White-sided dolphin. *See* Pacific white-sided dolphin
White-tailed deer, 78, 130, 139, 143-47
White-tailed jack rabbit, 19, 21
Wildcat. *See* Bobcat; Lynx
Wolf, 43, 75, 76, 77-80, 83, 88, 111, 151, 154, 155, 169, 170, 172
Wolverine, 43, 73, 78, 109-11, 165
Wood bison, 159, 162
Wood rats, 49. *See also* Bushy-tailed wood rat
Woodchuck, 23, 26-27, 28, 30, 99, 117, 121
Woodland caribou, 154

Yellow-bellied marmot, 28, 30
Yellow-pine chipmunk, 24, 25
Young, care of. *See* Parental care

*Zalophus californianus. See* California sea lion
Zapopidae. *See* Jumping mice
Ziphiidae. *See* Beaked whales